A Celebration of John Evelyn

Engraved by Swaine after Nanteüil.

A CELEBRATION OF JOHN EVELYN

PROCEEDINGS OF A CONFERENCE TO MARK THE TERCENTENARY OF HIS DEATH

Edited by
MAVIS BATEY

SURREY GARDENS TRUST

First published in Great Britain in 2007 by
SURREY GARDENS TRUST

DESIGN BY MARION WOODWARD

ILLUSTRATIONS RESEARCH BY
MARION WOODWARD
BERYL SAICH
CAROL HILL
BRENDA LEWIS

British Library Cataloguing in Publication Data.
A catalogue record for this book is available
from the British Library

ISBN 978-0-9540630-2-3

Printed and bound by Wellington Press Ltd.
Units 7 & 8 Four Seasons Crescent
Kimpton Road, Sutton, Surrey SM3 9QR

Contents

Acknowledgements

The Publications Committee of the Surrey Gardens Trust would like to thank all those individuals and institutions that have allowed us to reproduce illustrations from their archives and we are especially grateful to those, as well as the authors, that have generously allowed us to use their images free of charge, namely: Chris Beetles, British Library, British Museum, Mary Day, Dumbarton Oaks, East Sussex Record Office, Houghton Library at Harvard, Richard Ingle, Missouri Botanical Garden, National Library of Australia, National Maritime Museum, Orleans House Gallery at Twickenham, Shell Archive, Surrey History Service, Mark Davison, author of *Surrey In the Hurricane*, Stephen Torode and David Wrightson.

In addition, the Illustrations Research Team would like to thank all those individuals who have been extremely helpful and patient in providing guidance into the intricacies of image and permission ordering, notably: Guy de la Bédoyère; Melissa Dalziel, Principal Library Assistant, Imaging Services, at the Bodleian Library, University of Oxford; Alex Edouard of the Bridgeman Art Library; Trudy Southern of British Library Customer Services; Virginia Ennor, Copyright Administrator, Photography and Imaging, at the British Museum; Louisa Dave, Gallery Images, Courtauld Institute Gallery; Jennifer Nash and Philip Bye at East Sussex Record Office; Linda Campbell, Secretary to the Earl of Euston, Euston Hall; Jeremy Smith at the Guildhall Library, London; Thomas A. Horrocks, Associate Librarian for Collections at the Houghton Library at Harvard; Nikki Braunton, Picture Library Researcher, Museum of London; Natasha Mulder, Image Sales Coordinator at the Science & Society Picture Library; Fred Keusenkothen, Imaging Laboratory Coordinator at Missouri Botanical Garden; Margaret Daly at the National Gallery Picture Library; Sylvia Carr, Pictures Branch, and Loui Seselja, Photographer, National Library of Australia; Emma Butterfield, Picture Librarian, and Bernard Horrocks, Copyright Officer, at the National Portrait Gallery; Mark De Novellis, Curator of Exhibitions & Collections, Educations & Children's Services, Orleans House Gallery, Twickenham; Karen Lawson, Senior Picture Library Assistant at the Royal Collection; Veneta Bullen, Scala Archives, London, and Michael Casartelli, Purchasing Manager, Victoria & Albert Museum, for the introduction to Scala Archives; Nicky Balfour, Manager of the Shell Art Collection; Joanna Ling and Katherine Marshall of Sotheby's Picture Library; Kurt Johnson, Intern, Office of Research Librarian, at the Harry Ransom Humanities Center, University of Texas at Austin; Anna Smith, Picture Researcher at the Wellcome Trust; and Sue Walker, Assistant Librarian at the Lewis Walpole Library, Yale University. Our thanks also to Charlotte Brooks and Lucy Waitt, Picture Librarians at the Royal Horticultural Society Lindley Library, for their help in finding suitable illustrative material, to Julian Pooley and all the helpful staff at Surrey History Centre and to Don Josey for his fresh eyes at the final proofreading stage.

The Surrey Gardens Trust is also extremely grateful to the Marc Fitch Fund, the Paul Mellon Centre for Studies in British Art and N.M. Rothschild & Sons Ltd for providing funding to enable us to enhance this book with a large number of illustrations and to Jane Crawley, Gilly Drummond, Michael Symes and Colin Treen for acting as referees on our behalf.

Whilst every effort has been made to trace and contact the copyright holders of the illustrations used, we apologise to those whom we have not contacted or whose illustrations have been incorrectly credited. Please bring to our attention any discrepancies and they will be acknowledged in any future editions.

Finally, the Surrey Gardens Trust would like to express its sincere thanks to Mavis Batey for agreeing to act as Academic Editor for this book. Her matchless enthusiasm and far-ranging wealth of knowledge have added significantly to the final product.

Note from the Editors

We have had some difficulty over whether to use *Sylva* or *Silva* when referring to John Evelyn's renowned work on trees. The first three editions dated 1664, 1670 and 1679 were entitled *Sylva* but, by 1706, the publishers had adopted *Silva* as being the correct Latin spelling and later editions appeared with that title. However, the problem is compounded by the fact that later editors, e.g. Nisbet who reprinted the 1706 edition in 1908, reverted to the original spelling, *Sylva*. Our policy has been to use the version of the title as it appears on the title page of the edition actually being cited in references or credited in a caption but, in general discussion, we have used *Sylva* as being the most familiar title and the one chosen by Evelyn.

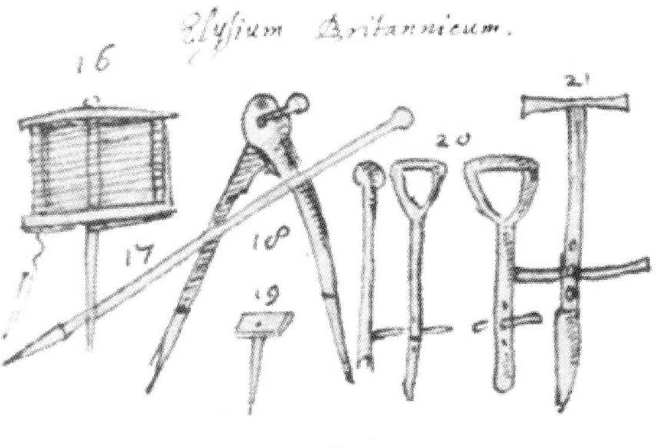

INTRODUCTION

The tercentenary of John Evelyn's death in 2006 provided the Surrey Gardens Trust with a unique opportunity to celebrate his contribution to trees, plants and gardens. Born at Wotton, near Dorking, Surrey always had an important place in his heart. His *Diary* gives significant descriptions of the places he visited and the people he knew and is probably better known today than his book on trees, *Sylva, or a discourse of Forest Trees*, for which he was most widely regarded at the end of his life. At my suggestion, the Council of the Surrey Gardens Trust agreed that the celebration would take the form of a two-day conference to be commemorated by the publication of the speakers' papers in this volume.

Organising a two-day conference is quite a task for volunteers and a sub-committee was formed, under the chairmanship of Jill Leggatt, comprising John Bridel, Bernice Hall, Carol Hill, Richard Ingle, Brenda Lewis and Sue White from the Surrey Gardens Trust and Robert Peel representing the Garden History Society. A list was drawn up of eminent Evelyn scholars who might speak; they were contacted and all agreed to lecture and donate their papers. The venue had to be Wotton House as it was Evelyn's birthplace; it is now a Hayley Conference Centre. The weekend of 22-23 April 2006 was blessed, after a long winter, with the first really fine weather of the year.

The Saturday morning lectures were given by Gillian Darley, Sally Jeffery and Frances Harris and, after lunch, David Jacques addressed delegates in and on the garden at Wotton. They then went to Wotton parish church, where many of the Evelyn family are buried, and later to Albury for a visit by kind permission of Michael Baxter and the Albury Estate. Here Dominic Cole talked on the gardens and the delegates enjoyed afternoon tea in Albury Park House, thanks to Jennifer Whalley and her chef. The first day ended with a dinner at Wotton which was addressed by Patrick Evelyn, DL.

On Sunday morning, papers were read by Douglas Chambers, Maggie Campbell-Culver and Peter Brandon; Mavis Batey concluded the proceedings with a masterly summing-up and the conference delegates dispersed after lunch. Special thanks are due to Peter Brandon who stepped in at the last minute as Mark Laird was unable to attend. Mark kindly agreed that a version of the paper he intended to present should be included in this book.

Some one hundred and fifty people attended the conference and it was voted a great success. After the event the papers and some suggestions for illustrations were received and much discussion took place between Mavis Batey and Marion Woodward in an attempt to achieve balance and consistency. Further illustrations were researched by the Surrey Gardens Trust's Publications Committee from a wide variety of sources.

This book would not exist without the dedication and enthusiasm of Marion Woodward. Thanks to her artistic flair and attention to detail, it is the book you now hold in your hands. She has given much of her time and energy over the last year to producing this volume, for which the Surrey Gardens Trust is deeply in her debt.

Brenda J Lewis,
Company Secretary, Surrey Gardens Trust

SURREY GARDENS TRUST

The Surrey Gardens Trust is a registered educational charity, founded in 1991, to help identify, preserve and recreate the county's outstanding gardens and landscapes, large and small, and to increase awareness of this heritage. It is part of a countrywide network of Trusts and is financed by subscriptions, grants and donations.

For further information about the Trust, visit our website at:
www.surreygardenstrust.co.uk

FIG.1.1: JOHN EVELYN BY HENDRICK VAN DER BORCHT, 1641

1

MEMOIRES FOR MY GRAND-SON JOHN EVELYN'S LEGACY

GILLIAN DARLEY

In his mid eighties, John Evelyn began to note down advice for young John Evelyn, his grandson and heir. At twenty-two he showed 'all those good & laudable qualitys & Inclynations which nourish & support my Hopes'. His grandfather, while handing on an inventory of their possessions at Wotton and giving out advice on estate management, from cleaning gutters to keeping poachers out of the warrens, was mapping out a life for him which was, essentially, the sum of his own experience as well as his pleasure in 'living for ingenuity'.[1]

Aged ten, Jack, as he was known, was sent to Eton – an option also offered to Evelyn who preferred to continue to live with his kindly step-grandmother and attend Southover Free School, in Lewes [Fig.1.2]. Then he went up to Oxford, perhaps surprisingly (given Evelyn's poor opinion of his old college), to Balliol. Jack's education had been, therefore, 'something above that of most ordinary Country Gents ...' – Evelyn abhorred the boring, vacuous life (gathering moss as John Aubrey put it) of the country estate, the drinking, gambling and field sports which had been the undoing of his nephews, and his own son, bringing the Evelyn estate to the very brink of ruin.

FIG.1.2: SOUTHOVER HOUSE, LEWES, IN 1826

FIG.1.3: FRONTISPIECE TO *THE HISTORY OF THE ROYAL SOCIETY OF LONDON* BY THOMAS SPRAT SHOWING WILLIAM BROUNCKER, 2ND VISCOUNT BROUNCKER, KING CHARLES II AND FRANCIS BACON, VISCOUNT ST. ALBAN.
ETCHING BY WENCESLAUS HOLLAR, AFTER JOHN EVELYN, DATED 1667

Thus education was an effort to arrive at 'something usefull as well as for merely entertainement of time'. He advised 'a constant and setl'd method' – with spare time well used – for nobody was too great or too busy to learn. Evelyn provided headings (he was fond of lists and catalogues); those of law, history, art, natural sciences and languages, ancient and modern. He emphasised particularly what he called practical mathematics which Francis Bacon had said, 'sharpens and settles the Judgement' and an area in which he felt deficient.

Lord Chancellor Bacon, Evelyn's 'noble Verulam' and author of (amongst many other books and essays) the *Advancement of Learning*, the *Novum Organum* (from the year of Evelyn's own birth, 1620) and *New Atlantis*, was 'a Lawyer, Judge, Privy-Consellor, & in perpetual buisinesse'. Fully engaged in a world beyond the laboratory or college study, he argued for the importance of natural philosophy (the new science) within the state. The Royal Society motto, which Evelyn helped suggest, was *Nullius in Verba* [Fig.1.3], 'on no man's word' – implying that the Aristotelian world of the 'ancients' must accommodate broader, and newer, lines of enquiry. A more recent Lord Chancellor, Edward Hyde, the 1st Earl of Clarendon had espoused the same principles and been instrumental in setting up the society. As he laid out his advice for Jack, Evelyn was currently reading his (posthumously published) three volume *History of the Rebellion* (1702-04) with intense interest and admiration.

FIG.1.4: SAMUEL PEPYS (1632 - 1703)

In Evelyn's life, his own education had been provided by travel, and with that the personal contacts and experience which could release what he called 'the noblest facultys of ... Intellectual Man'. But the *Memoires* do not mention anything about going abroad. All too aware that the Evelyn family future now rested on Jack's shoulders alone, he did not want to risk his life. Though he was adamant that Jack stay on English shores, Evelyn's friend of forty years Samuel Pepys [Fig.1.4] disagreed strongly. The memorable, too brief, journey he took with his wife, Elizabeth, in the autumn of 1669 was planned with Evelyn's help and had provided 'a degree of satisfaction and solid usefullnesse that has stuck by mee through the whole Cours of my Life and Businesse since'. So he ventures a counter argument '... since you aske it, I cannot but in faythfullnesse tell you that were hee mine, and (if it were possible) ten times more valuable than he is' he advised sending him abroad for some months, 'through Holland and Flanders to Paris' for '... those whose knowledge has been widen'd and refin'd by Travell' are at a great advantage over those 'whose Observation have been stinted to the narrow Practice of theyr own country.' Pepys, at the end of his life, envied his nephew, with Holland still ahead of him on his long travels.[2]

THE
THIRD BOOKE
Of Drawing, Limming, Colouring, Painting, and Graving.

By IOHN BATE.

LONDON,
Printed by *Thomas Harper* for *Ralph Mab.*
1635.

FIG.1.5: TITLE PAGE TO THE THIRD BOOK IN JOHN BATE'S *THE MYSTERIES OF NATURE AND ART*, PUBLISHED IN LONDON 1635

Evelyn's five (broken) years in Paris (between 1646 and 1652) were his real education, but the two months in Holland and Flanders in 1642 opened his eyes to the world at large. From his maternal grandfather, John Stansfield, he had learned that England was a gateway to Europe, the North Sea and English Channel not barriers but trade routes. Stansfield's business was shipping, his fleet came out of Newhaven – one vessel was named after Evelyn's mother, the *Eleanor Stansfield*. His cargoes included iron, wheat and lead, going to ports as distant as Marseilles and even Newfoundland.[3] Although he had been long dead when Evelyn headed for Holland – at much the same age as Jack now was – the imprint was made; John Evelyn was essentially a grandson of tradesmen and merchants (his paternal grandfather, Richard Evelyn, being the founder of the family gunpowder business) not just a member of the provincial landed gentry.

Before Evelyn left, his sister Jane asked for a portrait. The painter was Hendrick Van der Borcht, a member of the artistic court around Thomas

Howard, 2nd Earl of Arundel, who had become a neighbour of the Evelyns when he acquired an estate at Albury, only a few miles from Wotton, in the mid 1630s. Evelyn sat to Van der Borcht, also an art dealer and print publisher, at his studio in Arundel House [Fig.1.1]. Wenceslaus Hollar worked in an adjoining studio. This was the youth his family already called 'the Philosopher' and whose motto was *omnia explorate, meliora retinete* – 'look into everything, keep the best'.

He set off in summer 1641 with excellent recommendations for his forthcoming journey. A little album of drawings by Hollar, with several of Amsterdam and Leiden, includes an inscription with Evelyn's name (and the date 1641).[4] With advice from both Van der Borcht and Hollar he was unlikely to miss much. As well as becoming an avid print collector (a taste he first indulged in Paris) Evelyn was an enthusiastic draughtsman (despite his father's continual remonstrations that 'limning' wasted his time) and in 1649 etched two suites of landscapes, one imaginary, one Italian. The equipment was still at Wotton – he told Jack he would find a cabinet with tools for etching, engraving, boxes of colours, pencils and palette [Fig.1.5].

The excitement that Evelyn felt in Holland, seeing the visible rewards of successful international trade and commerce and the pride of a tolerant civic society, which accorded respect to all its citizens, jumps from the pages of the diary.[5] He visited wool halls, corn exchanges, weighing stations, the great printing houses as well as that cornucopia of marvels, the Dutch East India company headquarters in Amsterdam [Fig.1.6]. He admired the fashion for classical architecture which was

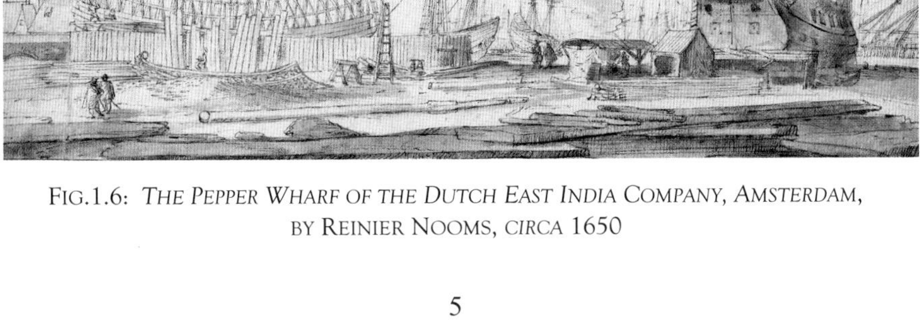

FIG.1.6: *THE PEPPER WHARF OF THE DUTCH EAST INDIA COMPANY, AMSTERDAM,* BY REINIER NOOMS, *CIRCA* 1650

replacing the old northern Renaissance style and saw how an ancient university had embraced the pursuit of new learning, visiting the Anatomy Theatre and Botanic Garden in Leiden [Fig.1.7]. He never forgot the neat brick buildings, the lime trees planted along canal and roadside, the ambitious largely man-made landscape in which the water had been engineered into compliance. He travelled home with Arundel, who was to be his mentor and guided him in northern Italy. But in the United Provinces he saw the wide benefits of shared knowledge, to the benefit of state and citizen – a more open and liberal version of Bacon's model Solomon's House, as described in his utopian *New Atlantis.*

All his life Evelyn was thirsty for guidance, whether on the page or in person. Choose a friend, he suggested to Jack, someone in whom you have complete trust and whom you can 'open your Mind to; such a one is

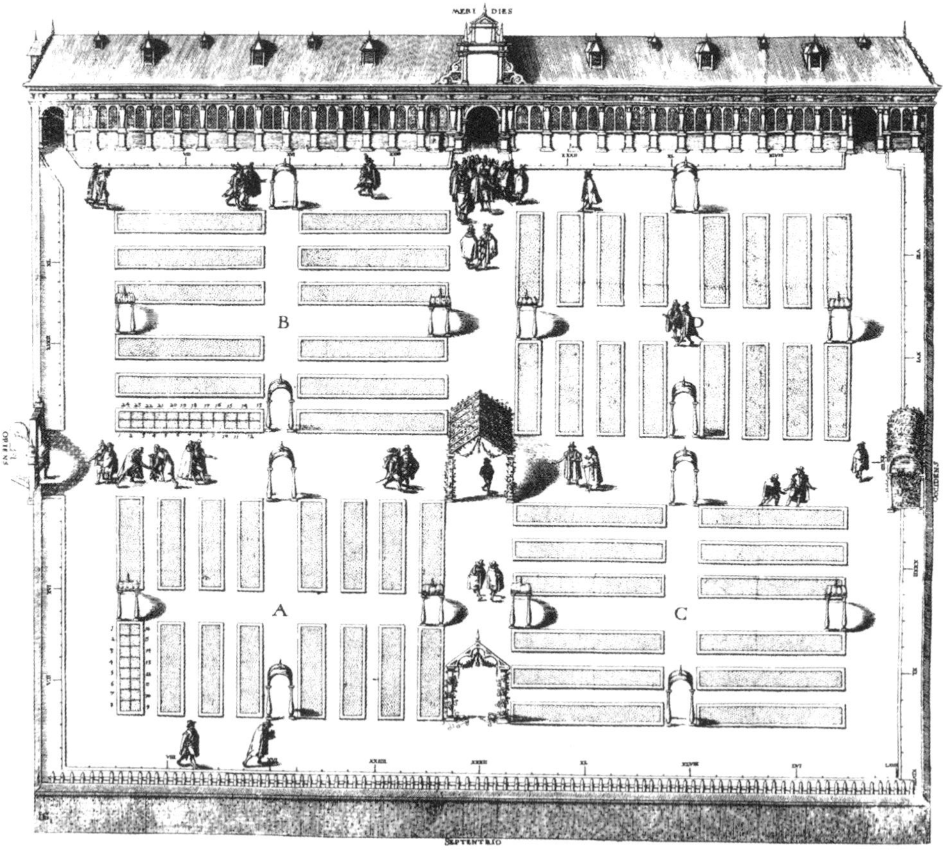

FIG.1.7: THE BOTANIC GARDEN AT LEIDEN, THE EARLIEST SCIENTIFIC GARDEN IN WESTERN EUROPE. SEVENTEENTH-CENTURY ENGRAVING

FIG. 1.8: SIR RICHARD BROWNE
BY R. NANTEUIL, PARIS 1650

indeede to be esteem'd an inestimable Treasure & hard to be found'. Then he should also have a 'pious and learned Divine' who will advise on matters of conscience – a role played in his life first by the royal chaplain John Earle in Paris, then by Jeremy Taylor and later by Thomas Tenison, since 1694 the Archbishop of Canterbury. In Paris he had met distinguished English churchmen and literary figures in exile – especially the poet Edmund Waller and the linguist and translator Richard Fanshawe as well as his (soon to be) father-in-law, Richard Browne, the king's Resident in France [Fig.1.8]. They helped guide him in his reading and collecting as in his own work of translation and publication. The presses of Europe were pouring out improved editions, new and better translations and more scholarly interpretations of almost every subject under the sun; Evelyn was drawn to print.

In Italy and France he met men who were publishing the works that were the talk of Europe: in Rome, Athanasius Kircher, the Jesuit polymath, and in Paris Thomas Hobbes and Kenelm Digby. Latin made the literate world a small place. He encouraged Jack to study medicine, for anatomy 'concerns your owne Body & its wonderfull structure, as it is a Microcosme.' In Padua, where Arundel had guided him to some of the most secret and loveliest corners of the city before they said their farewells, (he died the next year), he had pursued anatomical studies and explored the garden of 'simples' [Fig.1.10]. On leaving he bought a set of dessicated 'tables' showing the vascular system, which were finally engraved and published in the Royal Society's *Philosophical Transactions* in 1702.

FIG. 1.9: MARY EVELYN
BY R. NANTEUIL, PARIS 1650

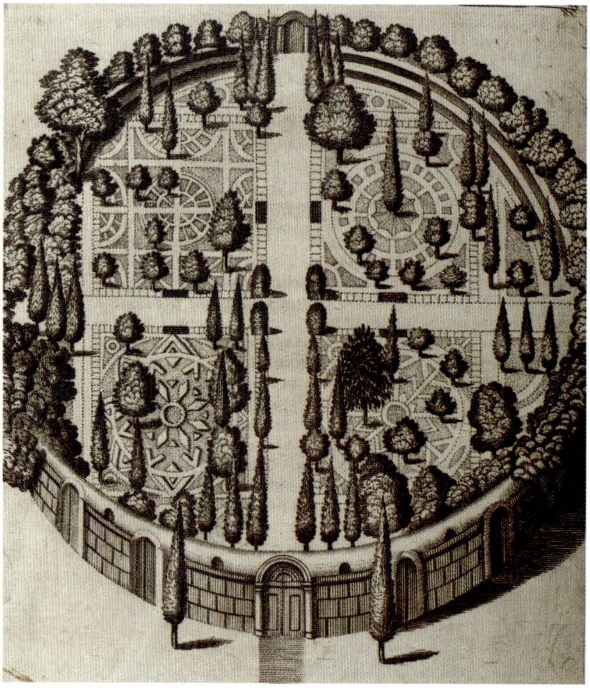

FIG. 1.10: THE BOTANIC GARDEN, PADUA, FROM J.P. TOMASINI, *GYMNASIUM PATAVINUM*, 1654

At Wotton, as at Sayes Court (his home until the mid 1690s), medicinal remedies (salves, ointments, spirits) and the skill of the 'still-house' were the woman's department. In 1705, while his grand-father was still compiling these notes, Jack married Anne Boscawen, Lord Godolphin's niece. She could do no better than to 'Imitate your Grand mother's Example of which I have to my comfort had about 60 years triall', a wife could be a help in many ways; 'you may safely confide in her'. There was a cabinet of curiosities at Wotton, 'natural and Artificial, collected casualy, usefull to Natural philosophy & contemplation of the work of God' amongst which a drawer of seeds, but such 'oddments, pieces & curiosities artificial' were there primarily to entertain 'virtuoso ladies' and the collection itself should become his wife's responsibility. After all, he pointed out, a discreet and faithful wife, such as Mary Evelyn [Fig.1.9] had always been, should be 'worthily Imploy'd and at home' – dividing her time between the still-room and the cabinet of curiosities.

Jack must read well and carefully, he told him. Evelyn advised him to keep a comprehensive commonplace book (a 'well digested Adversaria') so that he could take notes and not depend on memory alone. 'From such a Magazine one is inabled to speake or write upon any occasion' and so find some 'usefull subject to exercise your style in & to publish some Fruits of your studys, which cannot be don without Collections, no man being able to build anything whatever without the help [of] others ...' He discreetly touched on a sensitive issue, the disorder and inaccuracy of printing, the lack of copyright protection and the ambitions of literary men and women. Evelyn himself did not always remember to give credit where credit was due.

Evelyn's own ventures into print began with a translation (the best way to master a language, he advised) from La Mothe le Vayer (*Of Liberty and Servitude)* in 1649. He published his own *State of France* when he returned to England from Paris in early 1652. Then he engrossed himself in the books, friends and gardens he often celebrated as the ingredients of a contented life of withdrawal. He used the time at Sayes Court [Fig.1.11] to hone his horticultural and arboricultural expertise but once he began to venture beyond Deptford (in late 1654), he was irresistibly drawn into the world of natural philosophy, both in Oxford and London, through Dr John Wilkins at Wadham and Robert Boyle (the latter Evelyn's ideal of the godly scientist) as well as Samuel Hartlib and his west country

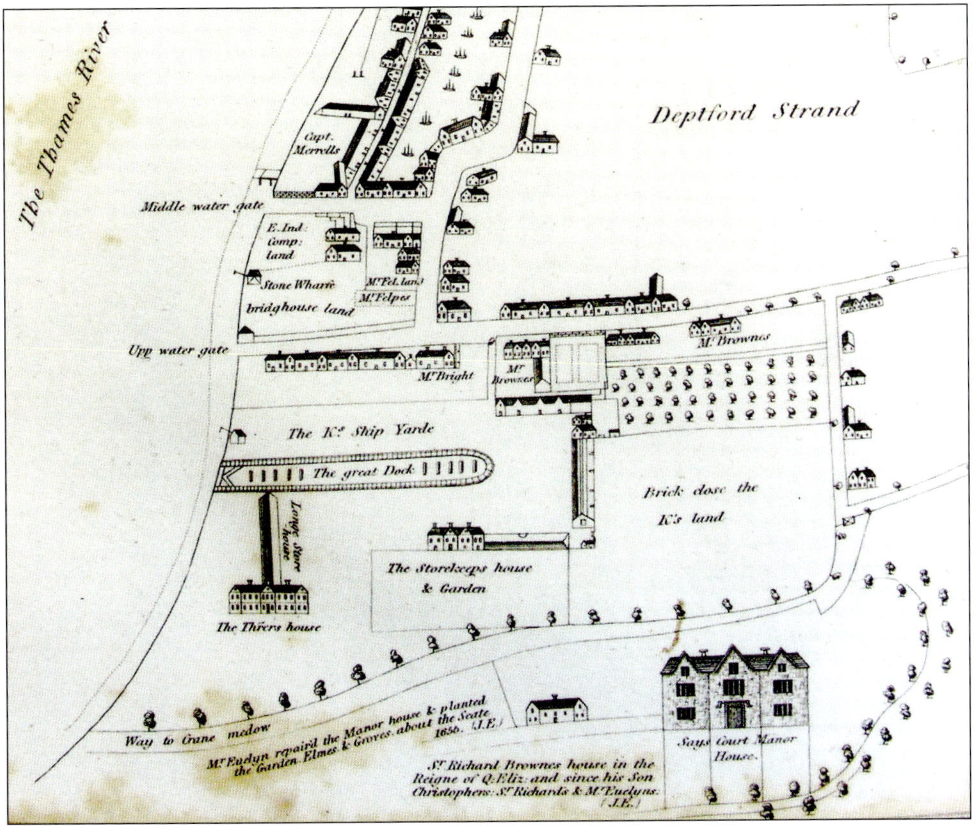

FIG.1.11: EXTRACT FROM *A MAP OF DEPTFORD, 1623,* FROM AN ORIGINAL PEN AND INK SKETCH WITH ADDITIONAL REMARKS BY JOHN EVELYN, ESQ.

FIG.1.12: AID TO PERSPECTIVE DRAWING FROM JOHN EVELYN'S *SCULPTURA*, 1662

correspondent, Dr John Beale. All encouraged him to build on his strengths, especially his knowledge of husbandry and horticulture, where he could become part of a network, in print and person. These were exchanges which were formalised once the Royal Society was chartered in 1662. Evelyn was a very early Fellow and a stalwart, lifelong supporter of what he saw to be the nearest equivalent to Bacon's Solomon's House – an enlightened forum for the free exchange of knowledge for the widest possible benefits.

In the *Memoires* Evelyn lists the instruments a virtuoso like himself owned; 'sphere [Fig.1.13], globe, perspectives [Fig.1.12], microscope, sexton, compasses, quadrants, rulers, tables, surveying instruments, optic glass, dials, needles, levels, glasses, barometer'. This mechanical treasure trove did its work; Sir John Evelyn, as he became in 1713, was elected a Fellow of the Royal Society in early 1723, nominated by Sir Hans Sloane, as well as becoming an early Fellow of the Society of Antiquaries, presenting the society with his grandfather's plan for rebuilding the City after the Fire of 1666 [Fig.1.14].

In the *Memoires* he told Jack that having 'written what is of such intrinsical value as to gaine universal applause', not to be timorous when venturing out again. He had begun translating Lucretius' *De Rerum Natura*, 'On the Nature of Things', into verse in Paris and published the first book in 1656 – hard-pressed to equate Epicureanism with his Christian faith. The list of stillborn literary projects in the

FIG.1.13: ARMILLARY SPHERE, 1554, MADE BY GIROLAMO DELLA VOLPAIA OF FLORENCE. EVELYN WOULD HAVE HAD A SIMPLER VERSION

Memoires is lengthy – he may not have regretted tricky ventures such as translations from French chemistry courses or the remaining five books of Lucretius, but he was saddened that he had never succeeded in completing a history of the stars in Latin, or putting together his thoughts on Spinoza, education, religion or politics. Above all, although he had translated, at Thomas Henshaw's suggestion, Nicolas de Bonnefons' *The French Gardiner*, he had failed to publish his own 'Elysium Britannicum' on which he had embarked in the late 1650s. He planned, he told his father-in-law, to present the 'Elysium' to Charles II –

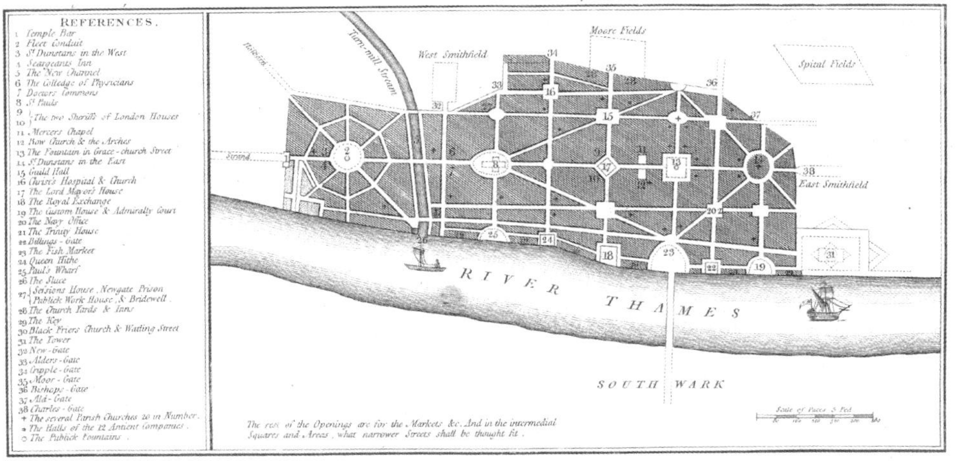

FIG.1.14: EIGHTEENTH-CENTURY PRINT OF JOHN EVELYN'S PLAN FOR REBUILDING THE CITY OF LONDON AFTER THE GREAT FIRE IN THE YEAR 1666

initially at the restoration, then for his coronation. But it was never finished – despite the help and urging of Sir Thomas Hanmer, Sir Thomas Browne, Dr Gaspar Needham (who marshalled a letter from assorted Oxford botanists and academics), Beale and Pepys, to whom he read, in late 1665, the 'discourse that he hath been many years and now is about, about Guardenage; which will be a most noble and pleasant piece'. When Robert Boyle brought John Rea's forthcoming volume to his attention, Evelyn thought it 'a very usefull and sincere booke' but since it only concerned 'the culture of flowers &c' no competition for his 'long since attempted designe of that entire subject'.[6] He never lost hope that the thousand pages of the 'Elysium Britannicum' with its hundreds of drawings, would be printed and handsomely illustrated, putting every other publication on the garden in the shade.[7]

FIG.1.15: TULIPA OCTAVIANI DEL PONT AND TULIPA ELEGANT FROM *HORTUS FLORIDUS*, PUBLISHED 1614-15. COLOURED ENGRAVING BY CRISPIN I DE PASSE (*CIRCA* 1565-1637)

Evelyn, in *Elysium Britannicum*, observes that the 'Florists ... make such a profession of selling and making gaine by their beauties' and 'have sold their curiosities of this nature at such prodigious and excessive rates, that it has bin thought expedient by the state [of Holland] that no man should give above a thousand *guilders* for a *Tulip* ... for to that intemperance and raritie were men arrived even in this innocent diversion.'

Even though Evelyn never left England again after 1652, the world that he created in the 'Elysium' brought Europe back to life, conjuring up the 'affections and rapture extraordinary' of Alps and Appenines or the 'natural and horrid Falls' at Tivoli and Frascati. Such natural features were exaggerated in these great Italian gardens, thanks to boundless expenditure. He treasured, too, memories of the royal terraces and noble gardens along the Seine. He paid tribute to the tulip [Fig.1.15] ('even Saloman in all his glory was not cloathed like one of them') and lauded (and borrowed from, with attribution) from his horticultural heroes ranging from Pierre Morin to Thomas Hanmer. Everyone who could bring back information (or better still plants or seeds) was asked to do so. Friends and relations, whether interested or not, performed their errands dutifully, and material came from France, Italy and Spain, as well as the Oxford Physic Garden and, through friendly strangers, from as far away as Virginia. But neither did Evelyn forget to celebrate what the English did best, their bowling-greens (or lawns as we know them) 'singular to the English nation above all others in the world'. The 'Elysium' kept pace with expanding knowledge of the natural world, but in the chaos of his own notes rather than in the light of the printed page. His final publication, *Acetaria* (1699), was just an extracted fragment (as was the immensely popular *Kalendarium Hortense*) [Fig.1.16] for he had been waylaid, as he put it, by 'the unsatiable coveting to exhaust all that could or should be said upon every head'.

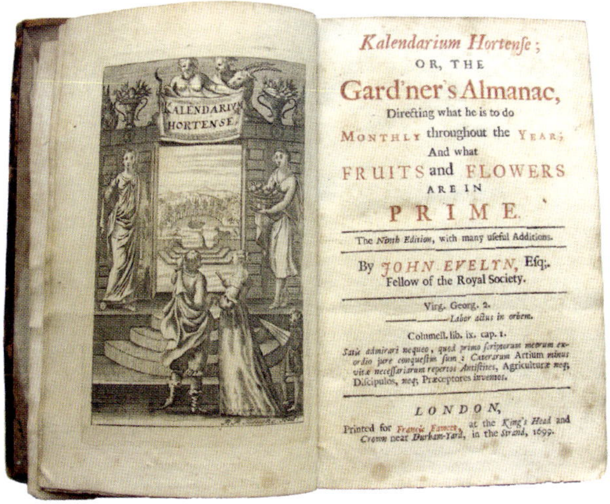

FIG.1.16: FRONTISPIECE AND TITLE PAGE FROM JOHN EVELYN'S *KALENDARIUM HORTENSE*, NINTH EDITION, 1699

Easier to finish were his translations of French titles on libraries, architecture and painting as well as his innovatory work on print-making, *Sculptura* (1662) – all projects begun in Paris and completed in Deptford, though none published until after the Restoration. In his pamphlet to greet the new king, *Fumifugium* (1661) [Fig.1.17], Evelyn could put his horticultural knowledge to a purpose, proposing a scented buffer zone, a 'green belt', between London and its industries. In *Sylva* (1664), the first official publication of the Royal Society, he was the author of an essentially practical volume.

So, although Mary Evelyn had occasionally been worried by the expenditure on the Sayes Court garden, Evelyn felt happy to advocate at Wotton the laying out of 'Groves and Walks and other innocent Amenitys' to his heir as an enjoyable way of preserving what is already 'planted and Improvable.' Within the garden, Jack must nurture fruit and orchard trees, especially wall

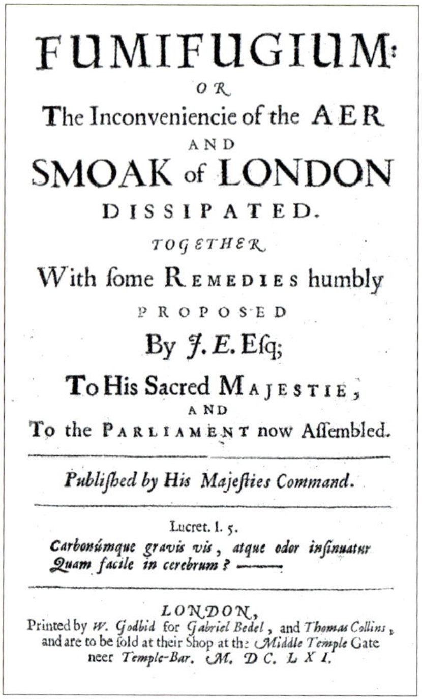
FUMIFUGIUM:
OR,
The Inconveniencie of the AER
AND
SMOAK of LONDON
DISSIPATED.
TOGETHER
With some REMEDIES humbly
PROPOSED
By *J. E.* Esq;
To His Sacred MAJESTIE,
AND
To the PARLIAMENT now Assembled.

Published by His Majesties Command.

Lucret. l. 5.
Carbonúmque gravis vis, atque odor insinuatur
Quam facile in cerebrum? ——

LONDON,
Printed by *W. Godbid* for *Gabriel Bedel*, and *Thomas Collins*, and are to be sold at their Shop at the *Middle Temple* Gate neer *Temple-Bar*. *M. DC. LXI.*

FIG. 1.17: TITLE PAGE TO *FUMIFUGIUM*

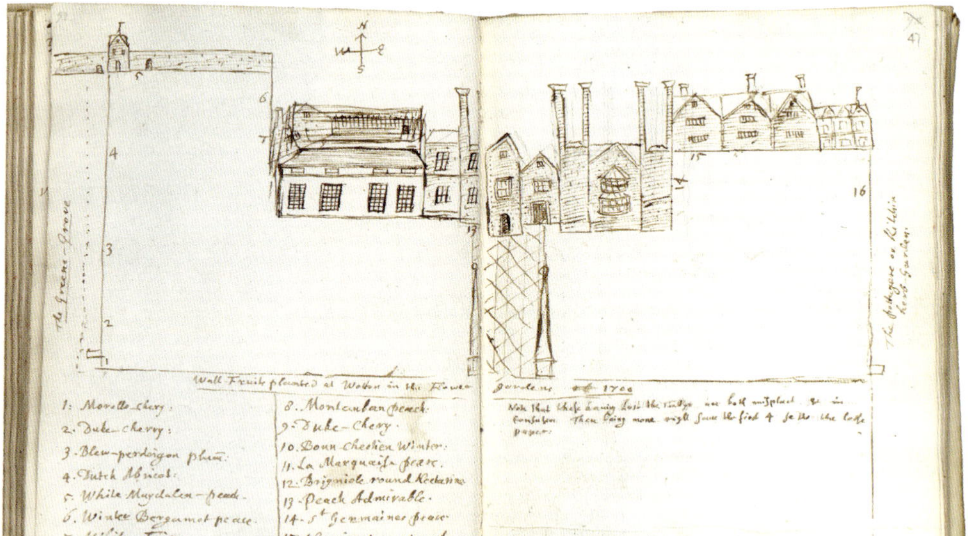

FIG.1.18: SKETCH OF THE HOUSE AND FLOWER GARDEN AT WOTTON WITH LIST OF FRUIT TREES PLANTED AGAINST THE WALL IN THE FLOWER GARDEN IN 1700

fruit (in 1700 Evelyn specified a complete new planting around the perimeter of the flower garden [Fig.1.18]).[8] If Jack planted anew (oak [Fig.1.20], ash [Fig.1.25] and elm, and 'frequent Copses' for which the soil was so suitable) Wotton would soon be profitable again, for this is 'in truth the onely best and proper Husbandry the Estate is capable of'. He was still working on the fourth edition of *Sylva* – forty years after the first.

Evelyn's own first-hand experience came from trial and error on the windblown site at Deptford and influenced the advice given liberally over

FIG.1.19: THE SEMI-CIRCULAR POOL AND FOUNTAIN AT ALBURY, SHOWING THE TUNNEL (CRYPTA) THROUGH THE HILLSIDE

the years to those who asked for his help. Henry Hyde, the 2nd Earl of Clarendon's Cornbury (and Swallowfield, his wife's estate) and Lord Arlington's Euston, were all parks laid out with his guidance. He designed two gardens for the future Duke of Norfolk, Henry Howard. One was a waterside 'public garden' in Norwich, laid out in the early 1660s and recorded admiringly, with its bowling green and diagonal paths, twenty years later.[9] Memorably, Evelyn remodelled Howard's garden at Albury [Fig.1.19] and made it as Italianate as was feasible on a

FIG.1.20: THE OAK TREE FROM SILVA, 1776 EDITION

Evelyn writes in *Sylva* (1679 edition): 'To enumerate now the incomparable Uses of this Wood, were needless: But so precious was the esteem of it, that of Old there was an express Law ... concerning the very gathering of Acorns, though they should be found fallen into another mans Ground.' He goes on to list some of its many uses: 'Houses, and Ships, Cities, and Navies are built with it ... Shingles, Pales, Lathes, Coopers ware, ... Water-works, Piles, ... Poles, Cudgels and Walking-staffs ...'

FIG.1.21: SIR CHRISTOPHER WREN
FROM AN ORIGINAL BY
SIR GODFREY KNELLER, 1711

Surrey hillside. William Cobbett found it looking splendid in the 1820s, though he did not spot the tunnel. Evelyn also offered help to those of his contacts building great houses in London, notably at Clarendon House and the trouble-some Lord John Berkeley's Berkeley House, on the western edge of London.

But he advised Jack to be frugal; who needed ostentatious buildings or furnishings? Evelyn had seen all too many families reeling from the extravagance of previous generations, including Howards and Clarendons. He was content, he told Pepys, to build (and draw up) just 'castles in the air'. But he also had practical suggestions; why did Jack not try burning brick and tile for sale (he had, perhaps, forgotten the failure of his own attempts to produce special vitrified brick for the post-fire rebuilding, using Dutch expertise at Deptford). When the pool at Wotton started to leak, old George suggested his brother ask Sir Christopher Wren [Fig.1.21], with whom he was then working closely on the Royal Naval Hospital at Greenwich [Fig.1.22], for his tips on waterproof repairs. Evelyn and

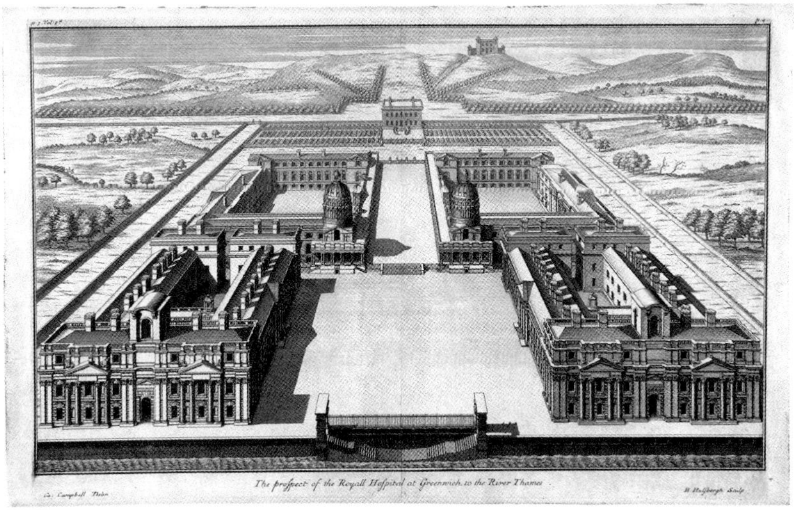

FIG. 1.22: THE ROYAL NAVAL HOSPITAL, GREENWICH.
ENGRAVING BY H. HULSBERGH AFTER C. CAMPBELL, 1715

Wren had, by then, known one another for fifty years and Evelyn's architectural knowledge was that of the expert amateur. In the 1690s he was busy copying out plates from a new French treatise, D'Aviler's *Cours d'Architecture*.

Having secured an education and put his affairs in order, Evelyn considered that a young man (even an eldest son) should be prepared for 'some laudable Imployment' such as the law, medicine, trade or the church. If a position in some office, including foreign service or court, can be obtained by 'Favour or purchase', then a father should help. But he firmly advised Jack not to become distracted, either in town or country, by 'large Acquaintance & Familiars' who will take up precious time. London is time consuming and expensive, especially with a wife and daughter, 'whole Time spent in play, or plays, impertinent Visits, double charges of Apparell, Mode, Gossipin, and for the most part un-acomptible Vanity'. It was a reminder of the debate about the active life versus that of contemplation in which, somewhat to his friends' surprise, Evelyn had argued for the engaged, active course as opposed to a withdrawn and contemplative existence.

Evelyn's own commitment to public service had been deflected by external events but no obstacles stood in Jack's path. If he intended to become a Member of Parliament it was essential to be 'dilligent, Impartial, studying how to serve the Country which chose you and to which you are in conscience to be accountable.' Suitable skills, 'dexterity of the pen, the Latine & modern Tongues &c' – especially French and Dutch at this date – allows 'young men to be usefull to the publique'. But he warns him not to take political office for reasons of vanity, popularity or 'the being caryed away by a faction or to serve a party'. Jack Evelyn was MP for Helston from 1708-10 – a brief spell compared to his great uncle George's many years in the Commons, at the end of which, Evelyn believed, he was the oldest member in the House. Jack was (briefly) joint postmaster-general but, as Sir John Evelyn Bt., became a commissioner of Customs for over forty years.

His grandfather's strong personal loyalties to men on both sides of what had become a party political divide ensured that he counted himself neither Whig nor Tory and he claimed he had turned down knighthoods more than once. His public service spanned four reigns, including two harrowing terms as a Commissioner to the Sick and Wounded and

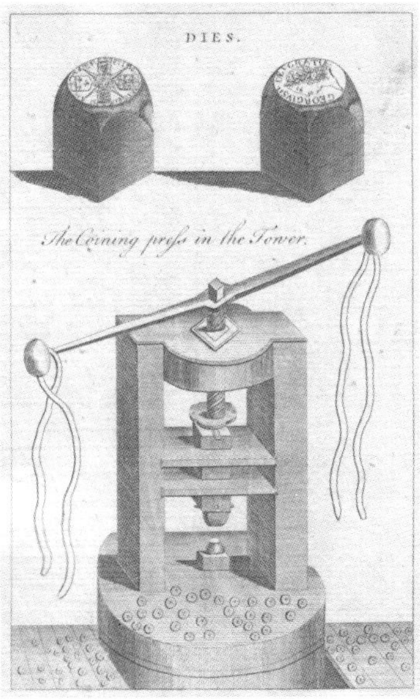

FIG. 1.23: A COIN MILL AT THE TOWER MINT IN LONDON, *CIRCA* 1761

Prisoners of War (in the 2nd and 3rd Dutch Wars), a period on the Council of Trade and Plantations and another as a Commissioner of the Privy Seal for James II. Finally, King William had appointed him Treasurer at Greenwich Hospital and, by the time he retired, Queen Anne was on the throne.

Finally but centrally, Jack must 'Give an Account of your Faith & choise of your Religion' which are founded 'upon principles solid & rational, and not because it is your Country's profession onely.' His grandfather had a long list of suitable theological reading matter. He also reminded Jack that the library needed additions, such as atlases and recent scientific transactions as well as French titles on natural history, and be purged of 'trifling Books'. He must keep a catalogue and note anything lent (even Pepys had mislaid some of Evelyn's family papers). There was a cabinet for medals, intaglios, coins, seals, modern coins, as might be expected from the author of *Numismata* (1697) and the man who had chosen the words from Virgil stamped onto the edge of the milled coinage [Fig.1.23] introduced in 1663, *Decus et Tutamen*, 'an ornament and a protection', words restored to modern pound coins. Satisfactorily enough, it was Jack Evelyn who later built a new library at Wotton.

The *Memoires* [Fig.1.24] with their emphasis on a questing life of the mind, underpinned by religion (a daily, not a weekly obligation), utter probity and a desire to be of service to the nation, show the breadth of Evelyn's ambition and the

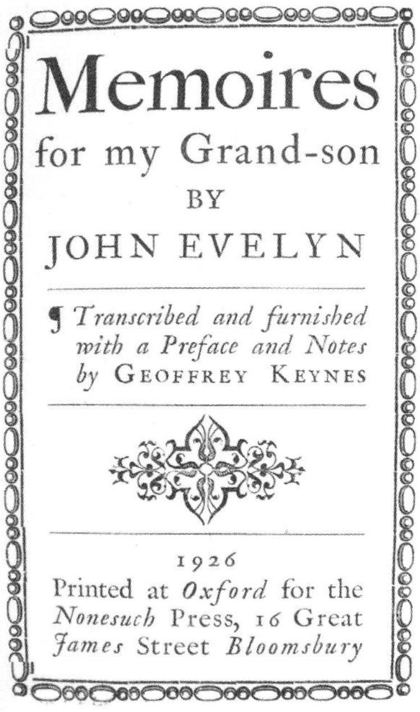
Memoires
for my Grand-son
BY
JOHN EVELYN

¶ *Transcribed and furnished with a Preface and Notes by* GEOFFREY KEYNES

1926
Printed at *Oxford* for the *Nonesuch* Press, 16 Great *James* Street *Bloomsbury*

FIG.1.24: TITLE PAGE OF *MEMOIRES FOR MY GRAND-SON*

FIG.1.25: THE ASH TREE FROM *SILVA*, 1801 EDITION

In *Sylva* (1679 edition) Evelyn stresses the numerous uses of the Ash tree: 'It serves the Souldier ... the Carpenter, Wheel-wrights, Cart-wright ... Cooper, Turner and Thatcher: Nothing like it for our Garden Palisad-hedges, Hop-yards, Poles, and Spars, Handles, Stocks for Tools, Spade-trees, &c. In sum, the Husbandman cannot be without the Ash for his Carts, Ladders, and other tackling. From the Pike, to the Plow, Spear and Bow ... in Peace, and War it is a Wood in highest request.'

strength of his personal morality and beliefs. Having always looked to exemplary men, he now presented himself and the fruit of his experience (as well as their shared family history) to Jack. This, he writes, is 'one of the most valuable legacys I can leave you' which will give 'an Early and universal prospect of all the Conserns you may find your selfe Ingaged in.' In 1657 William Rand dedicated his translation of Gassendi's *Life of Peiresc* to Evelyn, citing him as the most suitable man in England for the honour, in view of his many qualities but above all his 'learned Curiositie'. Horace Walpole saw Evelyn's life as following 'a course of inquiry, study, curiosity, instruction and benevolence'.[10] In his detailed and attractive prospectus for his grandson's own life of 'ingenuity' Evelyn drew a remarkable self portrait.

REFERENCES

1 The results were published as *Memoires for my Grand-son* by Geoffrey Keynes in a Nonesuch edition in 1926. All quotations not otherwise referenced, are to that volume.
2 *Particular Friends: The Correspondence of Samuel Pepys and John Evelyn*, ed. Guy de la Bédoyère (Woodbridge, 1997). See E14, SP to JE, 24 December 1701.
3 Colin Brent, *Pre-Georgian Lewes: c. 890-1714* (Lewes, 2004).
4 Now in the John Rylands Library, Manchester; Ryl. English MS 883.
5 Although written later, the account is his own. See vols. 1 (a later re-writing of the journey) and 2, pp. 29-75, *The Diary of John Evelyn*, ed. E.S. de Beer (Oxford, 1955).
6 R. Latham and W. Matthews eds., *The Diary of Samuel Pepys* (London 1971-1995), vol. 6, p. 289 (5 November 1665); Michael Hunter ed., *The Correspondence of Robert Boyle* (London, 2001), pp. 409-11, note p. 643, JE to RB, 23 November 1664.
7 See John Ingram ed. *Elysium Britannicum or The Royal Gardens* (Pennsylvania, 2001).
8 BL Add. MS 78349.
9 Thomas Baskerville's account (May 1681), BL Add. MS 70523.
10 Horace Walpole, *Catalogue of Engravers* (1763). Gillian Darley, 'John Evelyn's Norwich Garden', *Garden History*, vol. 34, no. 2 (2006), pp. 249-53.

GILLIAN DARLEY is a writer, biographer and broadcaster. She contributes to a wide range of publications, including the *London Review of Books*, the *Times Literary Supplement* and the *Literary Review*. She has been on the main committee of the Society for the Protection of Ancient Buildings since the early 1980s and chaired the society in the late 1990s. Her books include *Villages of Vision* (revised edition 2007), biographies of Octavia Hill and John Soane and *John Evelyn: Living for Ingenuity* published in 2006.

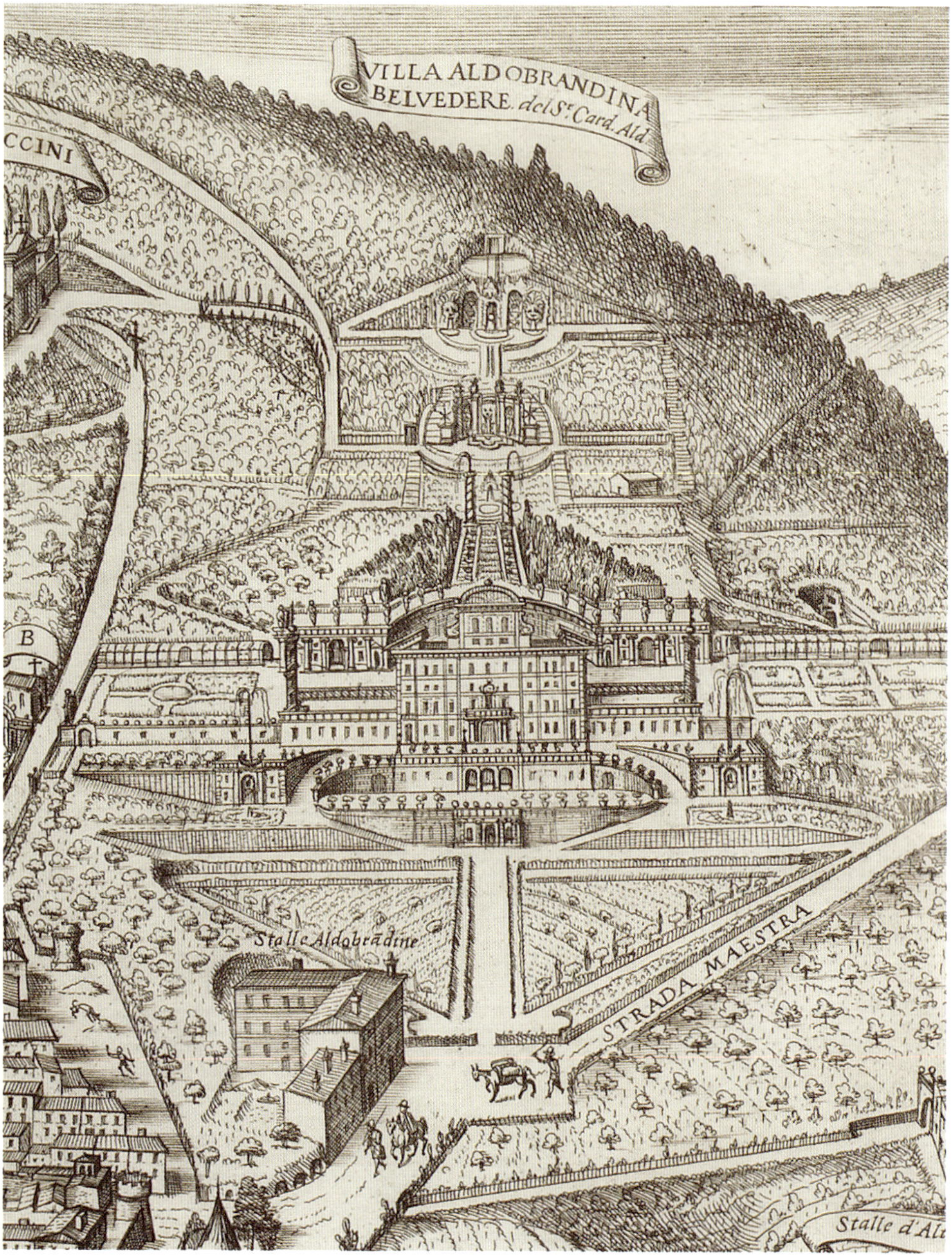

FIG.2.1: VILLA ALDOBRANDINI,
FROM A VIEW OF FRASCATI BY MATTHIAS GREUTER, 1620

2

THE WAY OF ITALIAN GARDENS

SALLY JEFFERY

John Evelyn left England for his three-year visit to France and Italy in 1643. In this paper, we will accompany him on his travels, seeing what he saw and hearing what he wrote. Were there gardens similar to those he knew at home? What impressed him most? How did his impressions influence him on his return when he commented on gardens in England and advised on their design?

He famously recorded his travels in a diary or 'Kalendarium'. He began this in note form in 1631, when he was eleven, but he only started writing seriously in the early 1660s, and some passages were probably written even later, in the 1680s. The account varies in the amount of detail given, with some material drawn from guidebooks, engravings, drawings, etc.[1] He owned copies of Claude de Varennes' *Voyage de France*, 1643 edition, and John Pflaumern's *Mercurius Italicus*.[2] He used others such as François Schott's *Itinerarium Italia*, of which he owned Latin, French and English editions, and those by William Thomas, John Raymond and Richard Lassels.[3] But many passages, particularly the ones on gardens, are his own, and very vivid.

When civil war threatened in 1641, he had briefly visited the Low Countries. His interest in gardens was already well established, and his diary comments announce themes which would be expanded later, such as, at Buitenhof, in the Hague, 'statues, Marbles, Grotts, Fountaines and artificiall Musique'.[4] In Amsterdam, he was attracted by 'the Streetes so exactly straite, even, & uniforme that nothing can be more pleasing, especially, being so frequently planted and shaded with the beautiful lime trees, which are set in rowes before every mans house, affording a very ravishing prospect'.[5] The fortifications at Willemstad and Antwerp [Fig.2.2] both planted with rows of trees, herald an interest in mounts and forts.[6]

Two years later, 'finding it impossible to evade the doing of very unhandsome things',[7] he obtained a licence to travel from the King and

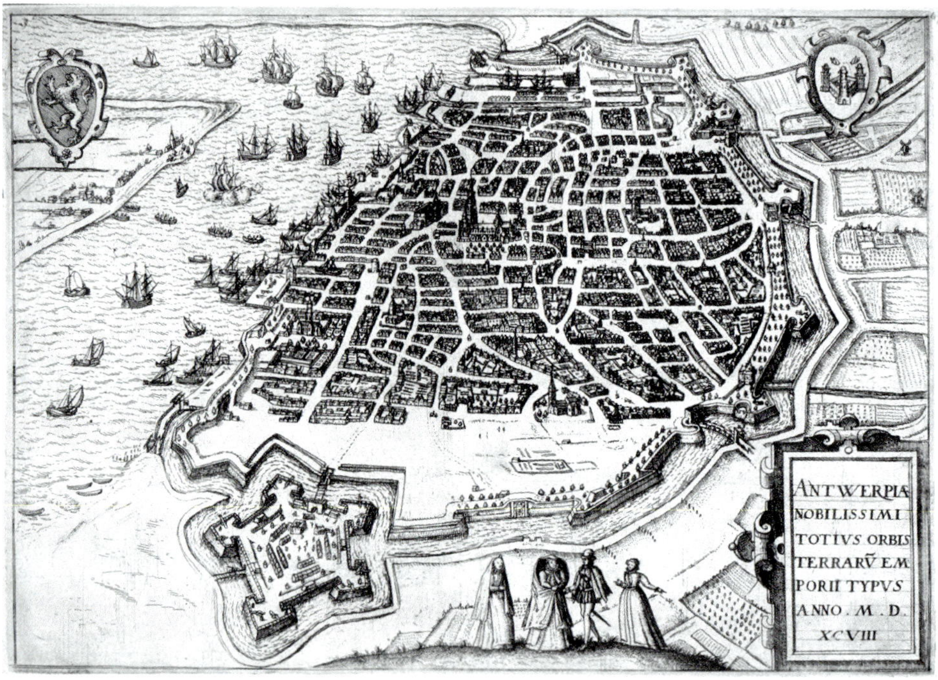

FIG.2.2: VIEW OF ANTWERP IN 1598 FROM 'TYPUS ANTWERPIAE'

FIG.2.3: THE POMEGRANATE (*PUNICA GRANATUM*). WATERCOLOUR BY CLAUDE AUBRIET (1665-1742)

left again, arriving at Calais on 11 November 1643, and staying in Paris that winter to visit the sights. He saw a number of gardens which clearly impressed him greatly, and gave him a foretaste of what he would see in Italy. Italian influence was strongly felt there, and ideas were flowing into France via publications and illustrations, and via travellers of all kinds.

At the Tuileries [Fig.2.4] he liked the plantations of trees, describing the walks of elms and mulberrys, and the hedges of pomegranate [Fig.2.3]. It was 'rarely contriv'd for Privacy, shade, company'. The echo in the semicircle at the bottom of the garden particularly pleased him 'redoubling the words so distinctly' and 'never without some faire Nymph singing to its grateful returnes',[8] and he used the example in 'Elysium Britannicum', with an explanatory sketch.[9]

Even though February and March were not the ideal season for garden visits, Evelyn and his companions took several journeys out to see palaces and gardens in the environs of Paris. An important excursion was to St. Germain-en-Laye. The road led them first to St. Cloud and then on to Cardinal Richelieu's Villa at Rueil, or Ruell as Evelyn calls it. Here he bestowed his greatest praise, saying that 'the Gardens ... are so magnificent, as I much doubt whether Italy have any exceeding it for all varietyes of Pleasure'. What he describes does indeed sound very Italian. Water jokes, 'fontaniere' to work them, basins, water flowing down steps 'like sheetes of transparent glasse', other cascades rolling 'with an astonishing noyse', 'jettos' and a grotto with a 'Yawning mouth of hell'.[10]

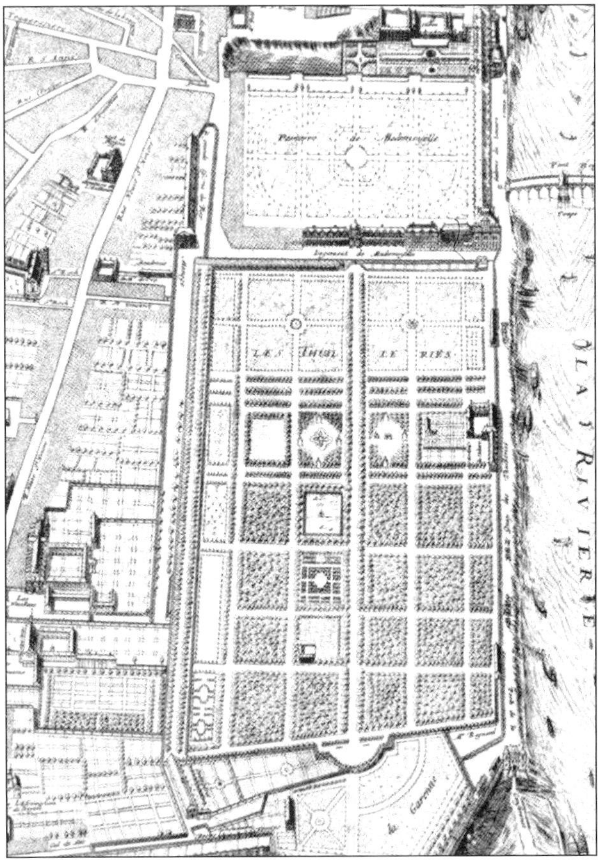

FIG.2.4: PLAN OF THE TUILERIES GARDENS, PARIS. DETAIL FROM JACQUES GOMBOUST, 'PLAN DE PARIS', 1652

At St. Germain itself [Fig. 2.5] they saw the gardens created for Henry IV from 1594:

> ... six incomparable Tarraces built of brick & stone descending in Cascads towards the river, & cut out of the natural hill, having under them goodley Gallerys vaulted, whereof 4 have subterranean Grotts & rocks, where is represented severall objects in manner of sceanes, & other motions by the force of Water ...[11]

FIG.2.5: VIEW OF ST. GERMAIN-EN-LAYE BY TOMMASO FRANCINI, 1614

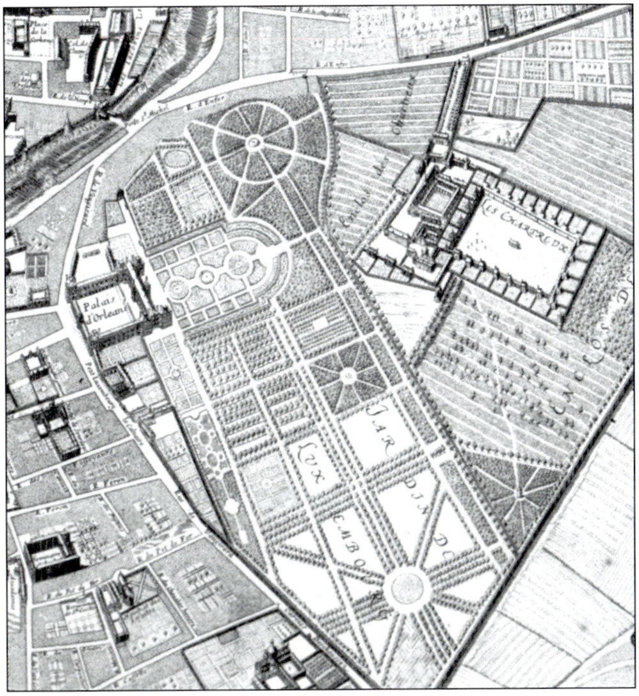

FIG.2.6: PLAN OF THE LUXEMBOURG GARDENS, PARIS. DETAIL FROM JACQUES GOMBOUST, 'PLAN DE PARIS', 1652

He was perhaps more taken with the garden of the Luxembourg Palace [Fig.2.6] than any other, judging by the length of his entry and his evident enthusiasm. 'I have been the larger in the description of this Paradise, for the extraordinary delight I have taken in those sweete retirements'.[12] The thing which pleased him most was the fact that the gardens were open, and were well used by 'Persons of quality, & Citizens, & strangers', together with 'Gallants & Ladys', 'melancholy Fryers', 'studious Scholars' and 'jolly Citizens', sitting or lying on the grass, running, jumping, playing bowls and ball, dancing and singing.

FIG.2.7: THE LIME TREE FROM *SILVA*, 1801 EDITION

There were numerous tree-lined walks – he noted elms and limes [Fig.2.7], and hornbeam hedges – and 'At the upper part (towards the Palace) is a grove of tall Elmes cutt into a Starr, every ray being a Walke whose center is a large fountaine.' When writing of this garden later, he criticised the fact that there was no long vista on the central axis, but this

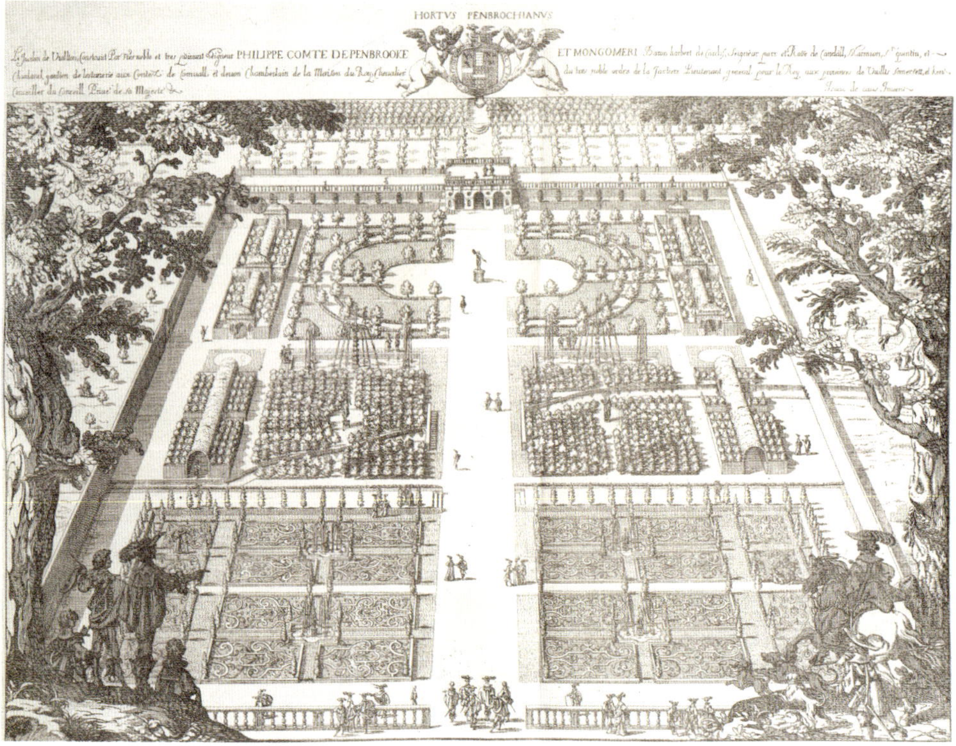

FIG.2.8 THE GARDEN OF WILTON HOUSE, WILTSHIRE,
FROM ISAAC DE CAUS, *LE JARDIN DE WILTON*, *CIRCA* 1630

was impossible because a Carthusian Convent owned the land behind, and the garden could extend only to the side.[13] Perhaps most surprising was what he said about the rest of the ground. It was 'made into severall enclosures (all hedge Worke or rowes of Trees) of whole fields, meadowes, boscages, some of them containing divers ackers'. Similar areas were described at Rueil and Richelieu. He appreciated the varied prospect of field and woodland in a semi-natural state.

This is not an element of French or Italian gardens with which we are particularly familiar, being more used to emphasis on the formal aspects, but it was a theme running through descriptions of seventeenth-century gardens at home and abroad. Evelyn knew Francis Bacon's 1625 essay 'Of Gardens', in which a certain 'Natural wildness' was celebrated in the farthest section of his 'Princely Garden'.[14] Henry Wotton, writing a year before Bacon, in his *Elements of Architecture*, spoke of gardens needing to be irregular, 'or at least cast into a very wilde Regularitie'.[15]

So in France, Evelyn was introduced to many new and exciting wonders – some specifically in the French taste, others directly reflecting what was to be seen in Italy. Was there anything to compare in England?

Almost all the gardens he visited were large, and many were rich in astonishing water features. He himself compared Fontainebleau to Hampton Court, and the only possible comparisons he could have made for St. Germain, Rueil and the Luxembourg would have been royal gardens, or those for people closely associated with the court circle of Charles I. One such, where hydraulics were used was Wilton House, near Salisbury [Fig.2.8], which he would have known by repute, if not personally.

It was rebuilt for Philip Herbert, 4th Earl of Pembroke, who had the magnificent garden, together with the house, remade '*al Italiano*'.[16] Inigo Jones, although too busy elsewhere to attend to all the details, nevertheless apparently supervised the work, which was executed by Isaac de Caus. The garden was laid out in 1632-5.

It famously contained an arcaded grotto [Fig.2.9] with three chambers, in the charge of Isaac de Caus, located under the terrace behind the main garden. Its watery interior was furnished with rocky, shelly walls, and mythical creatures. Hydraulic power worked tritons, singing birds, rainbows, figures weeping water on spectators, a table with a pipe spouting randomly to wet visitors and other water jokes.[17]

FIG.2.9: THE GROTTO AT WILTON HOUSE, WILTSHIRE, FROM ISAAC DE CAUS, *LE JARDIN DE WILTON*

The art of hydraulics, which was well-known to the ancients, underwent a revival in Italy, particularly in the second half of the sixteenth century. The skills were transferred to France by travelling hydraulic engineers such as Tommaso Francini, who went to work at St. Germain-en-Laye in 1598. Similarly, Salomon de Caus, a Frenchman and brother to Isaac, travelled to Italy to study, and published *Les Raisons des Forces mouvantes* in 1615, seeking to raise engineering skills to the level of art and recognising antique writings.

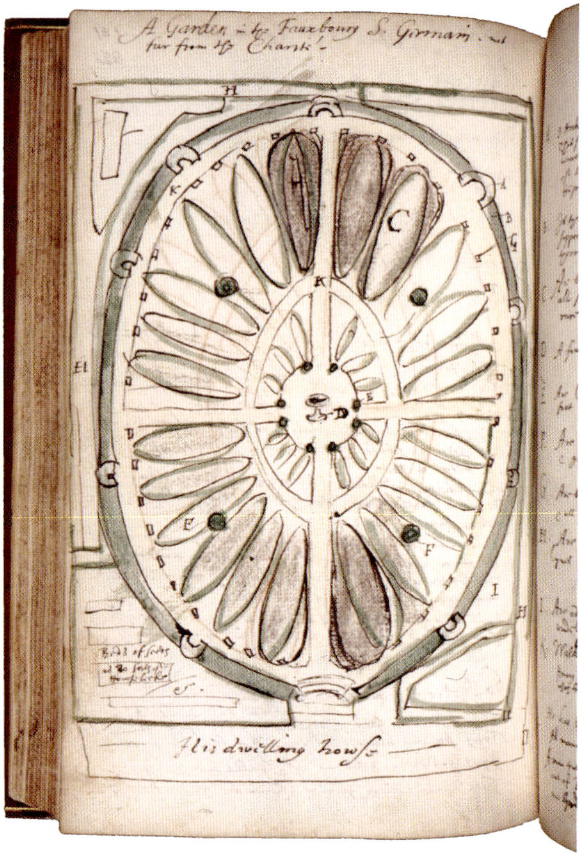

FIG.2.10: PLAN OF PIERRE MORIN'S GARDEN IN PARIS BY RICHARD SYMONDS, *CIRCA* 1649

He came to England to work for Queen Anne of Denmark at Greenwich and Somerset House and, briefly, for her son, Henry Prince of Wales, at Richmond. Evelyn may have seen his work in London. He was fascinated by hydraulics, and discussed and illustrated them at length in his manuscripts for 'Elysium Britannicum'.

Wilton garden was also exceptional in that it had embroidered plats or *parterres de broderie* – some of the earliest to appear in England – with raised terraces around them from which the patterns could be viewed. Evelyn particularly commented on the parterres he saw at St. Germain-en-Laye, the Luxembourg Gardens and elsewhere. This was a French fashion, grown out of ideas from Italy, and imported into England by means of engravings and French skill.

However, although such features were reserved for the largest gardens, the idea of a regular plan had been appreciated for some time, and was increasingly applied to gardens large and small. An ideal plan was illustrated by William Lawson in his book, *A New Orchard and Garden*, of 1618. This was a garden of modest size, perhaps two acres, which particularly emphasised the importance of symmetry.

In France, Evelyn did visit some smaller gardens. In April 1644 he was taken by a friend to see a garden belonging to a certain 'Monsieur Morine' – in fact Pierre Morin, one of the most important nurserymen in Paris. Morin was:

a person who from an ordinary Gardner, is arriv'd to be one of the most skillfull & Curious Persons of France ...: His Garden is of an exact Oval figure planted with Cypresse ...: The Tulips, Anemonies, Ranunculus's, Crocus's &c being of the most exquisite; were held for the rarest in the World, which constantly drew all the Virtuosi of that kind to his house during the season ...: He lived in a kind of Hermitage at one side of his Garden.[18]

Morin's oval garden was recorded by Richard Symonds, probably in 1649 [Fig.2.10].[19] Such an arrangement must have seemed less surprising to John Evelyn than the magnificences of St. Germain or the Luxembourg, and the concept of the Virtuoso Gardener who collected the rarities of the plant world would have been familiar to him. He became a great collector of plants in his own right when he returned from his continental travels and set about gardening at Sayes Court, Deptford.

Indeed the idea of 'an exact Oval figure planted with Cypresse' was in his mind when he designed the oval at Sayes Court [Fig.2.11],[20] which he referred to as his 'Morine Garden',[21] and M. Morin in his 'kind of

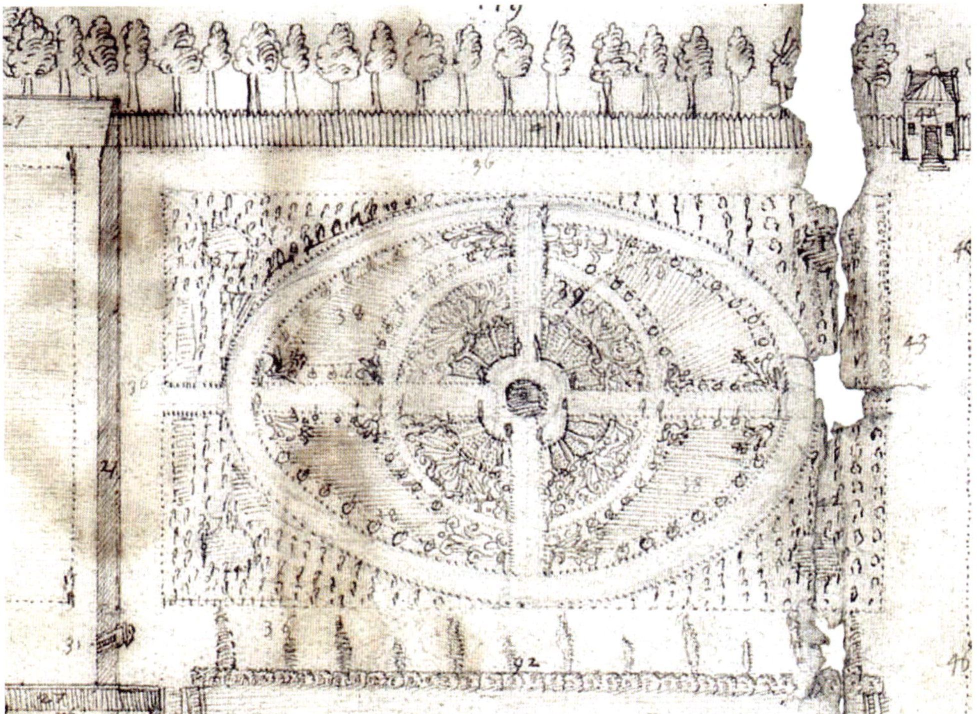

FIG.2.11: PLAN OF SAYES COURT, DEPTFORD, *CIRCA* 1653. DETAIL OF OVAL GARDEN

Hermitage' was not unlike Evelyn working in his 'elaboratorie' at Deptford.

Leaving Paris, Evelyn went down to the Mediterranean coast, and by boat to Genoa. After passing through Lucca, Florence and Siena, he spent several months in Rome, like most tourists. From Rome, he visited Naples. On the return journey northwards, he went through Florence again, Bologna, Ferrara, and Venice to Padua, and then on to Vicenza, Verona and Milan, crossing the Alps this time to travel across Switzerland and France to Paris. In all, he was in Italy a little over two and a half years, from October 1644 until May 1646.

Genoa was the first Italian city he saw. Fresh from France, he found it impressive and was astonished by the plants growing there.[22] Rubens' book, *I Palazzi di Genova* of 1622, which he quoted, illustrates the town palaces on the famous *Strada Nuova*. Because of the steep slopes, the gardens were often restricted in area, and were built on several different levels. In one garden in this wonderful street of palaces, Evelyn saw 'such huge Citrons hanging on the trees, applyd like our Abricotts to the Walles, that one would have believd incredible should have been supported by so weake branches'. Citrons (*Citrus medica*) are abundant in Genoa, because, as Evelyn observed they like the 'aire that proceedes from the Seas' [Fig.2.12].[23]

The huge terraced garden of the villa of the Principe Andrea Doria [Fig.2.13] was just outside the walls. Evelyn described the garden rising in terraces from the waterside up the hill, with fountains of white marble, 'an Aviary as Sir Fra: Bacon describes in his ... Essays, where in grow trees of more then 2 foote diameter, besides, Cypresse, Myrtils, Lentiscs & other rare shrubs, which serve to nestle & pearch all sorts of birds' and other gardens 'full of Orange-trees, Citrons & Pomegranads, Fountains, Grotts & Statues'.

Evelyn always had an eye for a prospect, and discussed the setting of various of the towns he saw and the approaches to them. For example, Lucca was 'a small but pretty Territorie ...: The Citty is neate, & well fortified, with noble and pleasant walkes of Trees on the Workes where the Gentry & Ladies use to take the aire'.[24] These tree-lined walks on the fortifications can still be seen.

FIG. 2.12: THE CITRON, VARIETY 'GIGANTEA'

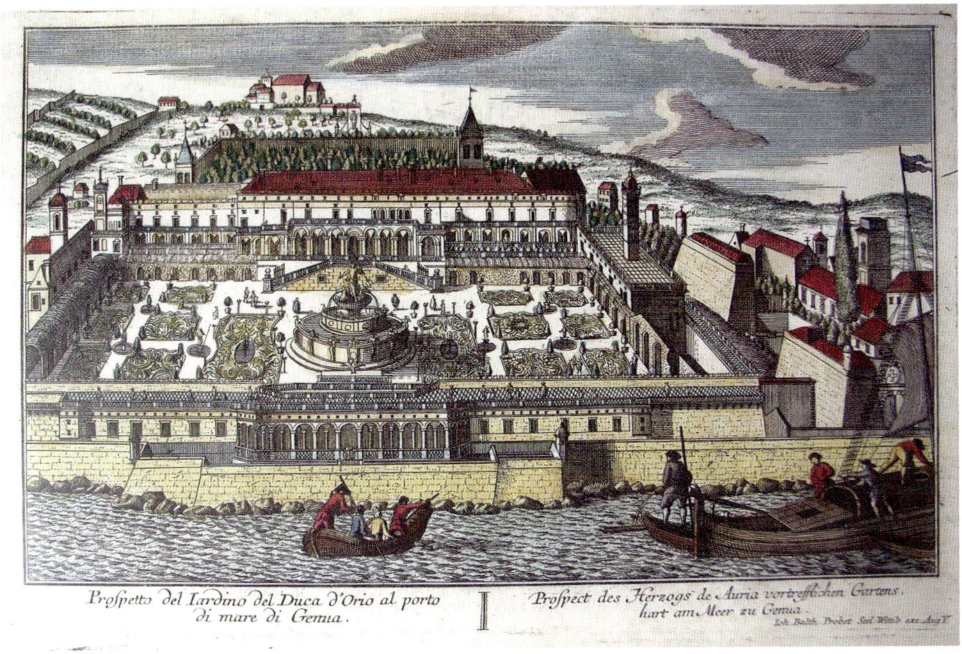

FIG.2.13: PALAZZO DEL PRINCIPE, THE VILLA OF ANDREA DORIA, GENOA

Some of Evelyn's most vivid impressions were formed during his time in Rome. Years later, in 1698, he recalled the 'many bright and happy moments' he had spent there with his friend Thomas Henshaw, visiting the 'villas and gardens of that glorious city'.[25] He used superlatives for his visit to Frascati to see the Villa Aldobrandini, constructed on a wonderful hillside site [Fig.2.1]. He wrote that 'for its elegance, situation & accommodation of plentifull water, Groves, Ascents & prospect' it surpassed 'the most delicious places that my eyes ever beheld'.[26]

FIG.2.14: VILLA ALDOBRANDINI, FRASCATI. WATER THEATRE

The exedra or water theatre [Fig.2.14] behind the villa was the centrepiece, with the water providing movement and power for the automata, sound effects, water tricks in the theatre and on the villa steps. There were 'many other devices to wett

FIG.2.15: THE WALK OF ONE HUNDRED FOUNTAINS AT VILLA D'ESTE, TIVOLI, BY GIOVANNI FRANCESCO VENTURINI (1650-1710)

the unwary spectators, so as one can hardly step without wetting to the skin', and the effects were spectacular. Villa Aldobrandini was one of three great gardens where Evelyn described the waterworks in detail.

FIG.2.16: VILLA D'ESTE, TIVOLI - FOUNTAIN OF VENUS FROM GIOVANNI BATTISTA FALDA, *LE FONTANE DI ROMA*

At Villa d'Este almost the whole description deals with jets, fountains, mirror pools and cascades. The Walk of the Hundred Fountains [Fig.2.15] was 'a long & spacious Walk, full of Fountaines, under which is historiz'd the whole Ovidian Metamorphosis in mezzo Relieva rarely sculptur'd'.[27] He said nothing here of the water jokes but, as usual, everyone got wet [Fig.2.16]. At the Medici villa of Pratolino, the garden was full of fountains and other curiosities. Evelyn and his friends were again 'well wash'd'.[28]

Pratolino was 'situate in a large meadow like an amphitheater'. Similarly, the Boboli Gardens in Florence were 'full of Variety, hills, dales, rocks, groves',[29] and in one garden in Genoa, the Palazzo Hieronymo del Negros, he was delighted by a grove 'furnish'd with artificial Sheepe, Shepheards & Wild beasts, so naturaly cut in a gre-stone ... that ... you would imagine your selfe in a Wilderness'.[30] All of this calls to mind the less formal areas of French gardens he had seen and admired earlier.

FIG.2.17: VILLA BORGHESE, ROME - NARCISSUS FOUNTAIN
FROM GIOVANNI BATTISTA FALDA, *LE FONTANE DI ROMA*

Tree-lined walks, often of evergreens, were another feature of Roman gardens which shaped his future views, reinforcing what he had seen in France and the Netherlands. They were everywhere, bordered with carefully clipped hedges, and set off with vases, statues and fountains. He noted particularly, at the Villa Borghese, the groves 'of Cypress and Lawrel, Pine, Myrtil, Olive &c.' [Fig. 2.17],[31] and at the Quirinale Palace, 'the glorious hedges of myrtle [Fig.2.18] above a man's height, others of Laurell, Oranges, nay of Ivy & Juniper'.[32]

FIG.2.18: MYRTLE (*MYRTUS COMMUNIS*) BY PIETRO ANDREA GREGORIO MATTHIOLI (1501-77)

Another feature of the gardens around Rome was the way terracing was used to create grand set pieces. Evelyn commented on it with great admiration. Thus, at Villa d'Este he spoke of the 'Ascent of the Palace', very evident in this seventeenth-century engraving [Fig.2.19]. Visitors used to enter from the bottom of the garden, which increased the dramatic effect.

On the same day as the visit to Tivoli, they also saw a spectacularly large Roman display of terracing – the ruinous Temple of Fortune of the second century AD at

FIG. 2.19: VIEW OF VILLA D'ESTE, TIVOLI, FROM GIOVANNI BATTISTA FALDA, *LE FONTANE DI ROMA*, 1670

Praeneste, or Palestrina [Fig.2.20] which he mentions almost in passing: 'And now we returne to Rome ...: by the way we were shew'd (at some distance) the Citty Preneste'.[33] It was one of the most significant sites and was probably already known to him by reputation. It consisted of a series of terraces and

FIG.2.20: VIEW OF THE ROMAN TEMPLE AT PALESTRINA SHOWING SOME OF THE TERRACING

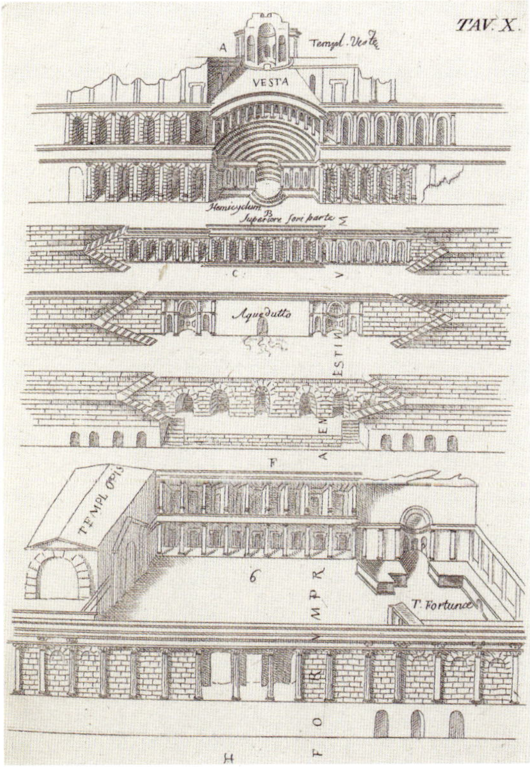

FIG.2.21: CONJECTURAL RECONSTRUCTION OF THE ROMAN TEMPLE AT PALESTRINA BY PIRRO LIGORIO, REDRAWN AND PUBLISHED BY G.B. CIPRIANI IN *MONUMENTI DI FABBRICHE ANTICHE*, 1796

ramps, surmounted by steps and, originally, a circular shrine. Pirro Ligorio, the designer of the Villa d'Este, explored the site in 1545 and drew inspiration from it for the Villa d'Este. He found it fascinating, and made a conjectural reconstruction drawing [Fig.2.21].[34]

Further classical associations were to be enjoyed around Naples. They went to see Vesuvius, 'that famous Vulcano', travelling on mules as far as possible and then on their hands and feet. They approached the summit with caution, and Evelyn recounts: 'I layd my selfe on my belly to looke over & into that most frightfull & terrible Vorago, a stupendious pit' with, in its

FIG.2.22: THE CRATER OF MOUNT VESUVIUS, DRAWN BY JOHN EVELYN, 1645, AND ETCHED BY HIM

centre, a 'hill shaped like a great browne loafe ... continualy vomiting a foggy exhalation, & ejecting huge stones'.[35] While in this position, he made a sketch, which he later etched as part of a set which he dedicated to Thomas Henshaw, his travelling companion and 'fellow eyewitness' [Fig.2.22].[36]

FIG.2.23: VIRGIL'S TOMB AND THE GROTTO OF POSILIPPO FROM JOHN RAYMOND, *IL MERCURIO ITALICO*, 1648

John Evelyn made drawings throughout his life. He records that at nine years old he 'tooke so extraordinary a fansy to drawing, and designing' that he could 'never after weane my inclinations from it'.[37] We hear of him making sketches or taking prospects in Italy and France, but very little has survived – just a few sheets of drawings and etchings.[38]

On 8 February 1645, they went with their guide to Baiae, just outside Naples, and were shown 'the Poet Virgils Sepulchre, erected on a very steepe rock, in forme of a Small rotonda' [Fig.2.23]. Afterwards, they went into the 'Mountain Pausilipo' as he called it, the 'noble, and altogether wonderfull Crypta ... cut through a rocky mountaine ... for neere three quarters of a mile'.[39] This clearly impressed him greatly, and was indeed famous, and visited by every tourist.

Having left Naples, and returned to Rome for a while, Evelyn headed north towards Venice and then down the famous Brenta Canal towards Padua. There, he went to see 'the Garden of Simples, rarely furnishd with plants, and gave order to the Gardner to make me a Collection of them for an hortus hyemalis'[40] – literally a winter garden, or dried collection. This was one of the oldest physic gardens, founded for teaching purposes by the University of Padua, and has remained on its site, little altered, since 1545 [Fig. 1.10]. We do not know exactly what Evelyn took back with him, but by that time there were at Padua both 'simples' – medicinally active plants – and exotics. He had already been to the Garden of Simples at Leiden,[41] and the Physic Garden at Pisa [Fig.2.24], which had a 'noble Palme tree' from which he took a long branch.[42]

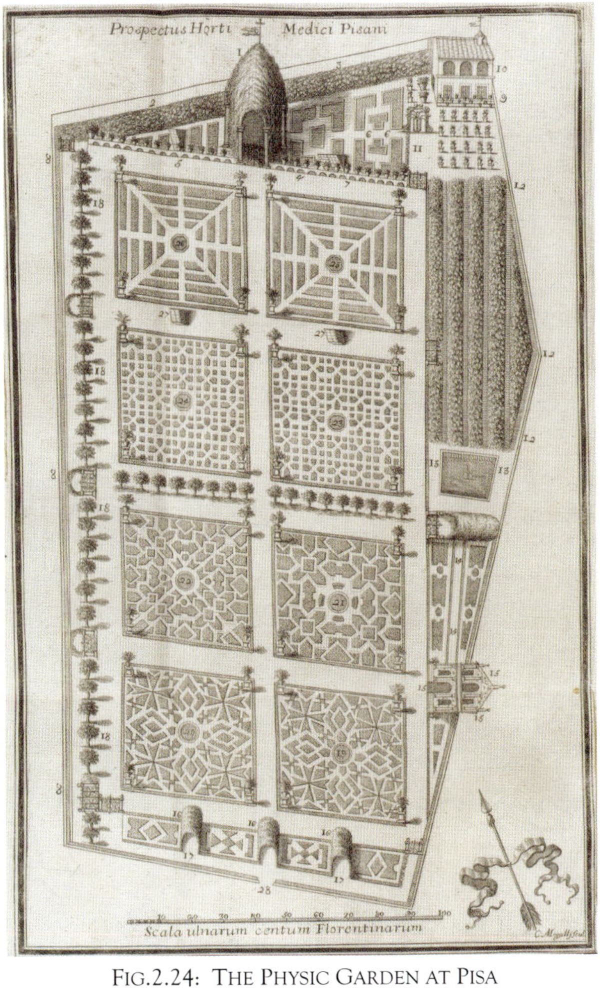

FIG.2.24: THE PHYSIC GARDEN AT PISA FROM MICHELE ANGELO TILLI, *CATALOGUS PLANTARUM HORTUS PISANI*, 1723

Evelyn signed the register at the University in Padua on 30 July 1645 intending to stay a few months to study physics and anatomy. Padua was important in Evelyn's journey for another reason. While there he resumed contact with Thomas Howard, 2nd Earl of Arundel, renowned collector, connoisseur of Palladian architecture, and a neighbour of the Evelyns at Albury in Surrey. He was an old acquaintance, and much

FIG.2.25: VIEW OF THE VALMARANA GARDENS, VICENZA, IN *CIRCA* 1770-80, ENGRAVED BY CRISTOFORO DALL'ACQUA

admired by Evelyn. Arundel's son, Henry Frederick Howard, and one of his grandsons, Henry Howard, were both fellow students of Evelyn's at Padua University that year.[43]

Before he left Padua, in April 1646, Evelyn breakfasted with the ageing Arundel, who received him in bed and wept over his family misfortunes, but sent him off with recommendations on what to see while travelling through northern Italy.[44] Arundel himself had famously visited the area over thirty years before in the company of Inigo Jones to study the works of Palladio. In Vicenza, he advised, 'The prime things to bee seene are the buildings ordered by Andrea Palladio the great Architect borne there, of wch. the prime are the Theatro with the Scene, which is the finest in Italy of that kinde'. Evelyn was not such a connoisseur of architecture as the Earl. Now, however, he paid homage

FIG.2.26: GIUSTI GARDENS, VERONA. VIEW OF THE LOWER GARDENS

to Palladio by visiting the Teatro Olimpico. He was well aware of the classical allusions projected by theatres, both as buildings and in landscapes.

In Vicenza, Arundel had also recommended the palace of 'Conte Teine' and the garden of Count Leonardo Valmarano.[45] This was separate from his palace, and situated just outside the walls on the road to Verona, and was completed by 1592.[46] Evelyn admired the maze and:

> his Cedrario or Conserve of Oranges ... set in exquisite order & ranges, & making a Canopie all the Way by their intermixing branches, for above 200 of my single paces, & which being full of fruite and blossomes, was one of the most delicious sights, that in my life I had seene.[47]

This appears towards the back of the engraving of the garden by Cristoforo Dall'Acqua [Fig.2.25].

FIG.2.27: GIUSTI GARDENS, VERONA. VIEW FROM THE TOP

At Verona, still following Arundel's advice, they saw the Roman Amphitheatre or 'Arena' and commented on 'the whole Citty bravely built'. 'In the Evening we saw Count Giusti's Villa, wher are Walkes cutt out of the maine rock, from whence we had the pleasant Prospect of Mantua, & Parma, though at great distance' [Figs.2.26 & 2.27].[48]

Evelyn and his travelling companions left Italy and crossed the Alps in about May 1646 and returned to Paris. He finally returned to England in 1652 to live at Sayes Court, Deptford.

Armed with all the impressions and information collected while he was away, he must have looked with new eyes at the gardens of his friends at home. Philip Packer, a fellow-student at the Middle Temple, replaced an old house at Groombridge Place, Kent, on a moated site between 1652 and 1674. The garden was established at the side of the house, rising up the hill, creating a terraced effect in three stages [Fig.2.28]. Evelyn visited in

FIG.2.28: GROOMBRIDGE PLACE. VIEW OF THE GARDENS

1674 and criticised it for its lack of prospect. He thought a better site could have been found up the hill, 'on a graceful ascent'.[49] A view was one of the most important elements in making a garden, and he approved, for example, of Wimbledon House, where he helped Lord Bristol to 'contrive the Garden after the moderne', and where there was 'a delicious Prospect'.[50]

The poet, Edmund Waller, having taken refuge in travel during the civil wars, spent some time in Italy with Evelyn. On his return he began to make a new house and garden at Hall Barn, near Beaconsfield, Buckinghamshire. Both house and garden survive, but it is difficult to be sure of the appearance of the garden in the mid-seventeenth century.[51] However, we know from a letter of 1724 that there were two 'pieces of water full of fish and handsomely planted & teraced on the sides', and extensive woodland, with groves of beeches, thick underwood, and walks.[52] The beeches [Fig.2.29] of the grove were tall by then, so must have been planted by Waller. There are no direct references to Evelyn visiting Hall Barn, but he remained in touch with Waller and it seems likely that the two discussed the project.[53] We can imagine Evelyn approving of the woodland and its walks.

Another garden which Evelyn knew and commented on was that of Charles Howard, a grandson of the 2nd Earl of Arundel, who had an extraordinary garden at the Deepdene, near Dorking, which was nestled into a natural recess in the hillside. It was much admired both for its

FIG.2.29 THE BEECH TREE FROM *SILVA*, 1801 EDITION

In *Sylva* (1679 edition), Evelyn tells us: 'The Beech serves for various Uses of the Housewife; with it the Turner makes Dishes, Trays, Rimbs for Buckets, and other Utensils, Trenchers, Dresser-boards, &c. likewise for the Wheeler, Joyner, for large Screws, and Upholster for Sellyes, Chairs, Stools, Bedsteads, &c. for the Bellows-maker, and Husbandman his Shovel and Spade-graffs; ... Not to omit even the very Shavings for the fining of Wines.'

situation as an 'Amphitheater Garden', and because of its owner's collections of dried plants and his laboratory.[54]

Evelyn's most powerful images of the Netherlands, France and Italy, judging by the diary, were of the importance of situation and prospect, exotic plantings, the extraordinary sculpting of the landscape into terraces, the spectacular waterworks, the avenues and walks, the theatre forms and antique settings, and also the less regular parts of some gardens. Many features mentioned in the diary reappear in the surviving manuscripts of 'Elysium Britannicum', so we can be sure of the things which most impressed him. Indeed, in the missing Book 3, there was to have been a section on 'the most Famous Gardens in the World, Ancient and Modern'.[55] In a letter to Sir Thomas Browne of 28 January 1658, Evelyn gave a full list of these, which included many he had seen himself, and said he intended to make it 'as particular in the descriptions as is possible' to be of use to travellers abroad.[56]

Evelyn was consulted frequently by friends and family about their gardens, but his involvement is often difficult to assess because specific information is not available. We do know, however, of a handful of gardens in which he was directly involved, and where his contribution was significant.

His garden at Deptford was, of course, his own invention. His recollection of Pierre Morin's in Paris, with its striking oval shape, was present in his mind during its formation, as we have seen. Sayes Court also bore witness to other things seen by its owner on his travels. It was approached by a stately lime avenue, and composed of a number of separate gardens, divided by hedges and walls into mostly geometric shapes. He had a 'Private Garden of choice flowers and Simples', and a grove of forest trees with little pavilions or arbours for contemplation. Other sections were planted as orchards and kitchen gardens. Still others were kept as fields, just as he had seen at the Luxembourg.

Another garden which he visited frequently and where he was consulted, was the small suburban plot of Sir Stephen and Lady Fox, adjoining Chiswick Villa. He was critical of its size, lack of water and situation, but must have approved of the huge greenhouse where exotics, including citrus, were grown.[57]

In 1664, he went down to Cornbury in Oxfordshire, the seat of the Earl of Clarendon to discuss the new buildings and gardens with

Clarendon's son, Lord Cornbury and with his friend the architect Hugh May. All three appear to have been involved in setting out the dramatic grand vista, and other walks in the woodlands.[58] The tree-lined walks first seen in the Netherlands, and then in France and Italy were surely a great influence on extensive forest gardens like Cornbury.

Evelyn loved the moated family house at Wotton as a child, and went back there as a young man. In January 1651, he tells us that he wrote to his brother from Paris about the garden[59] and, after his return to England in 1652, he visited him and observed the removing of some of the hillside to form the mount, filling up the moat, and levelling the area of the parterre and fountain.[60] His own rôle in all of this is not entirely clear, as Frances Harris shows, but what was created was a geometrically terraced mount, with a grotto set into it, with sculpture and fountains.[61]

This garden, therefore, had both waterworks and changes of level affording prospects. The mount is somewhat fortified in appearance. Evelyn could have told his brother about the various types of fortifications he had seen on his travels, including 'a Mount planted with Cypresses, representing a fortress'[62] at the Palazzo Medici in Rome [Fig.2.30], and the artificial island of Isola Bella on Lake Maggiore, which was only recently constructed when he visited [Fig.2.31]. He commented that 'the whole Iland is a Mount, ascended by severall Terraces'. These ideas may have helped to shape the mount at Wotton. However, more

FIG.2.30: VILLA MEDICI, ROME, FROM GIOVANNI BATTISTA FALDA, *LI GIARDINI DI ROMA*. THE CYPRESS COVERED MOUNT CAN BE SEEN IN THE DISTANCE TO THE LEFT OF THE PATH

significantly, there were examples of sculpted mounts in England which he and his brother might have known. The one in the garden at New College [Fig.2.32] was quite recently constructed when he was studying at Oxford.[63] So the features at Wotton may have derived from a mixture of sources, at home and abroad.

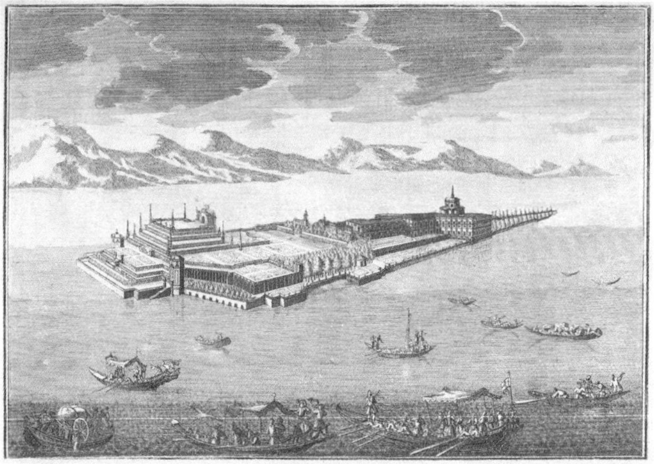

FIG.2.31: ISOLA BELLA, LAKE MAGGIORE, FROM MARC' ANTONIO DAL RE, *LE VILLE DI DELIZIE*, 1726

Albury was altogether different and was very much the product of his old friendship with the Earl of Arundel, and his interest in classical antiquity. Evelyn was directly involved with the design, which was almost entirely the result of impressions of both contemporary and classical sites gathered on his travels.

It had been acquired by the Earl in the 1630s, so the Howard family had long known the Evelyns as their Surrey neighbours. Arundel had undertaken considerable work there, which Evelyn probably saw when he visited from nearby Wotton before his departure for the Continent.[64] The hillside was then terraced into vineyards, with grottos set into the terraces. After Arundel's death in 1646, his grandson, Henry Howard, worked on a new design with Evelyn, his friend and fellow student in Padua. They discussed the terraced garden with

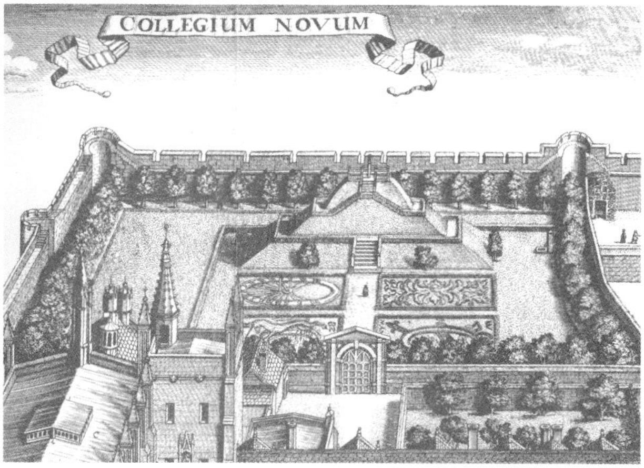

FIG.2.32: THE GARDEN OF NEW COLLEGE, OXFORD, IN 1675 FROM DAVID LOGGAN, *OXONIA ILLUSTRATA*

its canal, grass walks, bath house, exedra and tunnel through the hill. It is one of the most interesting Italianate gardens of its time, with strong echoes of the antique, and the dramatic hillside gardens seen by Evelyn around Rome.

In September 1667 Evelyn wrote: 'I accompanied Mr Howard to his Villa at Alburie, where I designed for him the plat for his Canale and Garden, with a Crypta thro the hill &c.' [Fig.2.33].[65] He later wrote that

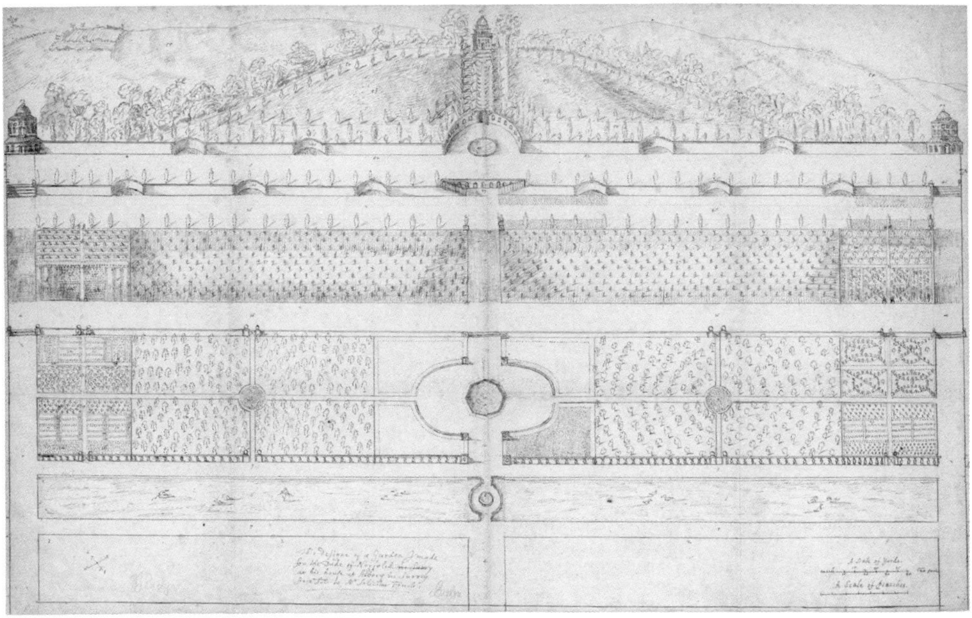

FIG.2.33: PROPOSED DESIGN FOR THE GARDEN AT ALBURY, SURREY, BY JOHN EVELYN, *CIRCA* 1667

it was 'don according to the Designe & plot I had made, with the Crypta through the mountaine in the parke, ... such a Pausilippe is no where in England besides'.[66]

He was clearly thinking of his visit to the famous tunnel near Virgil's tomb – a spot full of associations with the classical world and its literature. At Albury, the entrance to the pausillipo leads out of a niched exedra with a pool [Fig.1.19] – a Roman form seen by Evelyn at the Villa Aldobrandini and elsewhere. There was a grotto, or perhaps a bath-house, set into the terrace below, which used to have waterworks of some kind.

Most dramatic of all, however, were the two long terraces, connected by ramps [Fig.2.34]. He was recalling here not only the famous Roman Temple of Fortune at Palestrina, which he had seen on its hillside in 1645, but also the numerous references to its massive terraces made in Renaissance Italy and France – many also seen by him. He was using one of the most influential of classical designs as the basis for the garden of the Howards. He was right in saying that it was repeated 'no where in England besides'.

FIG.2.34: TERRACES AND RAMP AT ALBURY, SURREY

Evelyn was uniquely well-informed on Dutch, French and Italian gardens, as well as those in England. But did he consciously think of creating gardens inspired by his travels? Albury was exceptional in his work and was the garden most redolent of his Italian journey. Elsewhere, he drew on a whole range of sources at home and abroad, illustrating his love of variety, his fascination with tree planting, exotics and ingenious devices, his eye for an effective setting and his interest in less formal areas in gardens. Much of this emerged in his writings, and heralded the advent of such things later in the seventeenth century, and on into the eighteenth.

REFERENCES

1 Many of these sources are noted in E.S. de Beer (ed.), *The Diary of John Evelyn*, 6 vols. (Oxford, Clarendon Press, 1955). Subsequent references to this will be abbreviated to '*Diary*'.

2 The 1628 edition, purchased in Rome in the autumn of 1644, and now in the British Library. Giles Mandelbrote, 'John Evelyn and his Books', in *John Evelyn and his Milieu* (British Library, 2003), p. 92, fn. 20.

3 Details of the guidebooks he owned are in his 1687 library list, British Library, Add. MS 78632.

4 *Diary*, vol. 2, p. 41, 19 August 1641.

5 *Diary*, vol. 2, p. 46, *circa* 21-24 August 1641.

6 *Diary*, vol. 2, p. 60, 27 September 1641 and p. 67, 5 October 1641.

7 *Diary*, vol. 2, p. 82, 23 July 1643.

8 *Diary*, vol. 2, p. 106, February 1644.

9 John E. Ingram (ed.), *Elysium Britannicum or The Royal Gardens* (University of Pennsylvania Press, 2001), pp. 228-9.

10 *Diary*, vol. 2, pp. 109-110, 28 February 1644.

11 *Diary*, vol. 2, p. 111, 28 February 1644. This passage is derived from de Varennes.

12 *Diary*, vol. 2, pp. 129-131, 1 April 1644.

13 *Elysium Britannicum*, p. 126.

14 John Dixon Hunt and Peter Willis (eds.), *The Genius of the Place: The English Landscape Garden 1620-1820* (Paul Elek, 1975), p. 55.

15 Henry Wotton, *The Elements of Architecture* (1624), p. 109.

16 John Britton (ed.), *The Natural History of Wiltshire* (1847), pp. 83-4, 86-7. See Howard Colvin, 'The South Front of Wilton House', *Essays in English Architectural History* (Yale University Press, 1999), pp. 136-157.

17 Britton, op.cit.; *Wilton Garden* (n.d., *circa* 1630), pp. 22-4; Christopher Morris (ed.), *The Illustrated Journeys of Celia Fiennes 1685-circa1712* (Macdonald, 1982), pp. 38-9.

18 *Diary*, vol. 2, pp. 132-3, April 1644.

19 Prudence Leith-Ross, 'A Seventeenth-century Paris Garden', *Garden History*, vol. 21, no. 2 (1993), pp. 150-7.

20 Mark Laird, 'Parterre, Grove, and Flower Garden: European Horticulture and Planting Design in John Evelyn's Time', in Therese O'Malley and Joachim Wolschke-Bulmahn (eds.), *John Evelyn's 'Elysium Britannicum' and European Gardening* (Dumbarton Oaks, 1998), p. 181; Prudence Leith-Ross, 'The Garden of John Evelyn at Deptford', *Garden History*, vol. 25, no. 2 (1997), pp. 138-152.

21 Leith-Ross, op.cit. p. 153.

22 *Diary*, vol. 2, pp. 173-6, 16-17 October 1644.

23 *Elysium Britannicum*, p. 317.

24 *Diary*, vol. 2, p. 409, May 1645.

25 From William Bray (ed.), *The Diary and Correspondence of John Evelyn*, vol. 3, p. 376. See Douglas Chambers, ' "Elysium Britannicum not printed neere ready &c": The "Elysium Britannicum" in the Correspondence of John Evelyn' in Therese O'Malley

and Joachim Wolschke-Bulmahn (eds.), *John Evelyn's 'Elysium Britannicum' and European Gardening* (Dumbarton Oaks, 1998), p. 126.

26 *Diary*, vol. 2, pp. 392-3, 5 May 1645.

27 *Diary*, vol. 2, pp. 395-6, 6 May 1645.

28 *Diary*, vol. 2, pp. 418-9, May 1645.

29 *Diary*, vol. 2, p. 187, October 1644.

30 This was the garden of the Palace of Hieronymo del Negros. *Diary*, vol. 2, p. 173-4, 16-17 October 1644. It is now called the Villa di Negro-Durazzo-Rosazza, and a fragment survives as a public park.

31 *Diary*, vol. 2, p. 252, 17 November 1644.

32 *Diary*, vol. 2, p. 287, 29 November 1644.

33 *Diary*, vol. 2, p. 397, 6 May 1645.

34 Turin State Archive, cod.a.II.1.J.14, fol. 95r.

35 *Diary*, vol. 2, pp. 332-6, 7 February 1645.

36 Antony Griffiths, 'The Etchings of John Evelyn', in David Howarth (ed.), *Art and Patronage in the Caroline Courts* (Cambridge University Press, 1993), pp. 51-67.

37 *Diary*, vol. 2, p. 9, 1629.

38 There was 'A portfeuille of rede draughts of landscapes, buildings, surveys, & other papers drawne by J. Evelyn casually' in his 1687 library catalogue (see BL Add. MS 78632, and Antony Griffiths, 'John Evelyn and the Print', in *John Evelyn and his Milieu* (British Library, 2003), pp. 95-113.

39 *Diary*, vol. 2, pp. 337-8, 8 February 1645.

40 *Diary*, vol. 2, p. 466, 31 July 1645.

41 *Diary*, vol. 2, pp. 52-3, August 1641.

42 *Diary*, vol. 2, p. 181, 20 October 1644.

43 See Edward Chaney, 'Evelyn, Inigo Jones, and the Collector Earl of Arundel', in *John Evelyn and his Milieu* (British Library, 2003), pp. 37-60.

44 *Diary*, vol. 2, p. 479, late April 1646. For the Earl's recommendations, see John Martin Robinson (ed.), *Remembrances of Things Worth Seeing in Italy Given to John Evelyn 25 April 1646 by Thomas Howard, 14th Earl of Arundel* (Roxburghe Club, 1987).

45 *Diary*, vol. 2, p. 484, late April/May 1646. Both of these have been misidentified in past writings. See Chaney, op.cit., pp. 52-3.

46 His town house, the Palazzo Valmarano by Palladio, is on Corso Fogazzaro. A fragment of the garden survives, now called the Giardino Salvi.

47 *Elysium Britannicum*, p.325: 'Such a Cedrarie as this, we once saw in a Curious Garden at Vicenza, which the trees were all arched over in manner of a Galery, the pendulous fruite adorning the florid roofe, for neere a thousand paces, & which we beheld with a ravishing admiration'.

48 *Diary*, vol. 2, pp. 485-8, April/May 1646.

49 *Diary*, vol. 4, p. 40.

50 *Diary*, vol. 3, pp. 315-6, 17 February 1662.

51 The garden was modified by his grandson Harry Waller and his stepfather John Aislabie in the early eighteenth century.

52 British Library, Egmont MSS, letter of 9 August 1724, quoted in John Dixon Hunt and Peter Willis (eds.), *The Genius of the Place: The English Landscape Garden 1620-1820* (Paul Elek, 1975), p. 165.
53 *Diary*, vol. 3, p. 386, 26 October 1664. There are statues from the Arundel collection at Hall Barn, but it is unclear whether Waller placed them there, or whether they arrived later.
54 *Diary*, vol. 3, p. 154, early August 1655.
55 *Elysium Britannicum*, p. 23.
56 Geoffrey Keynes (ed.), *The Letters of Sir Thomas Browne* (Faber & Faber, 1931, reissued 1946), pp. 299-305.
57 *Diary*, vol. 4, p. 294, 30 October 1682.
58 *Diary*, vol. 3, p. 381, 17 October 1664.
59 *Diary*, vol. 3, p. 24, 3 January 1651.
60 *Diary*, vol. 3, pp. 60-1, 22 February 1652.
61 *Diary*, vol. 2, p. 551, 26 February 1649.
62 *Diary*, vol. 2, p. 231, 8 November 1644.
63 Work on this is recorded in 1584-5, 1616 and 1623, and it was complete by the 1640s. Eleanor Sinclair Rohde, *Oxford's College Gardens* (Jenkins, 1932), p. 41.
64 *Diary*, vol. 2, p. 77, 7 November 1641.
65 *Diary*, vol. 3, p. 496, 21 September 1667.
66 *Diary*, vol. 3, pp. 561-2, 23 September 1670.

SALLY JEFFERY is an architectural and garden historian and has lectured and written extensively on seventeenth- and eighteenth-century buildings and gardens. After obtaining her doctorate in 1986, she worked as a historian at the Corporation of London until 2002, researching the City's historic buildings and open spaces, and is now tutor and lecturer for the MA in Garden History at Birkbeck, University of London, and an independent scholar. She has a particular interest in English gardens of the seventeenth century.

FIG.3.1: *RHAMNUS ALATERNUS* OR THE ITALIAN BUCKTHORN
BY JEAN HENRI JAUME SAINT-HILAIRE, 1825

3

'MY MOST CHERISHED PLACE ON EARTH': JOHN EVELYN AND WOTTON

FRANCES HARRIS

John Evelyn loved Wotton, though he spent little of his life there. He told his wife that although they had chosen to live on her ancestral estate at Deptford, Wotton was still his 'most cherished place on earth'.[1] He loved its sense of rural seclusion, despite its proximity to London. He loved the landscape: the long chalk spine of the downs to the north and to the south the eminence of Leith Hill, from which spread out a great panorama of Kent and Sussex. He especially loved the pockets of thick greenery, rocky outcrops and small torrents which made a kind of miniature Salvator Rosa landscape [Fig.3.2]: 'such a solitude as I have never seen any place more horridly agreeable and romantick'.[2] Our knowledge of what Wotton looked like in the seventeenth century comes almost entirely from his drawings and engravings and his first garden experiments were made there. Having visited and admired Italian Renaissance gardens, he explicitly commented on Wotton's suitability for Italianate garden design.

FIG.3.2: *CAVE WITH WATERFALL*. OIL PAINTING BY SALVATOR ROSA (1615-73). FLORENCE, GALLERIA PALATINA

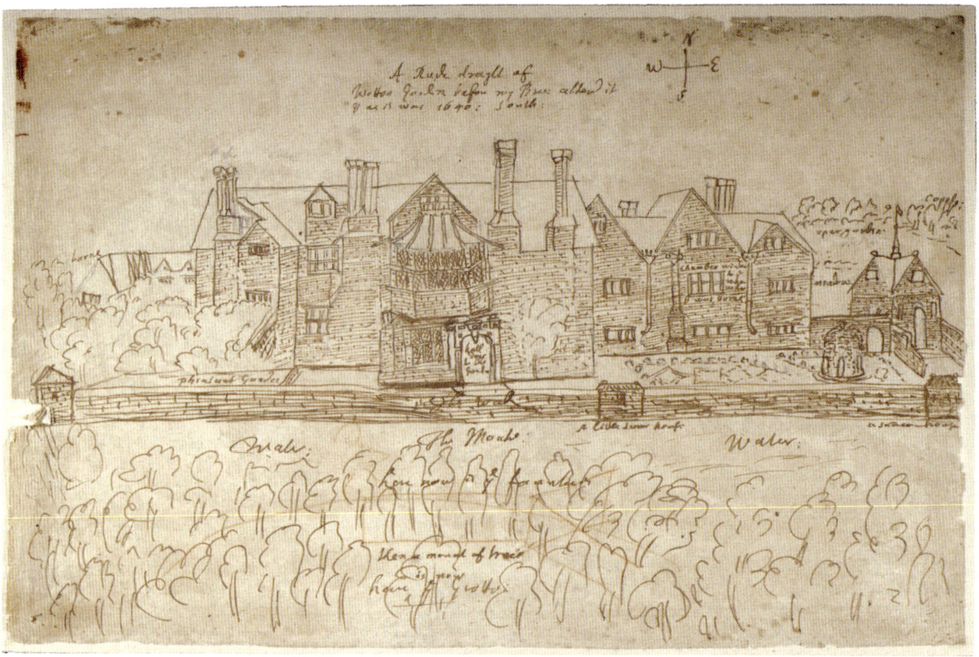

FIG.3.3: 'A RUDE DRAGHT OF WOTTON GARDEN BEFORE MY BRO: ALTERD IT & AS IT WAS 1640: SOUTH'. PEN AND INK DRAWING BY JOHN EVELYN. THE INSCRIPTION 'CHAMBER WINDOW TO THE ROOME WHER I WAS BORNE' IS ON SECOND BAY FROM THE RIGHT

So when, under the ownership of his elder brother George, who never travelled, Wotton acquired a full-scale Italianate garden in the 1650s it is natural to assume that John must have been responsible for the overall conception. But was he?

To one who had experienced the disruption of the Civil War and its aftermath, the century of stability and prosperity before it seemed a kind of golden age. John Evelyn harked back nostalgically to the 'ancient simplicity' of the way of life at Wotton in his father's time. All the records that survive of Richard Evelyn, the account-books and copious notes on the administration of his manors, in the well-formed hand which was unchanged from the time he inherited the estate in 1603 until his death in 1640, confirm his son's description of him as 'a thriving, neate, silent and methodical genius'.[3] They tell of his concern with every aspect of his estate and every means of improving its income. The Tudor manor house of Wotton, Evelyn wrote, was 'large and antient, suitable to those hospitable times'.[4] His drawings, one of them pin-pointing the window of the chamber in which he was born [Fig.3.3],[5] show it as rambling and irregular, with a moat and a pheasant garden on the south side. But this rural retreat was set in what was effectively an industrial landscape. To the early generations of the family the estate's combination of woods, variations of level and fast-flowing water did not suggest Salvator Rosa; they suggested fuel and power. 'Not far from my brothers house', Evelyn told John Aubrey years later, 'upon streams and ponds since fill'd up and drain'd, stood formerly many powder mills, erected by my ancestors ... in this parish were set up the first brass-mills for the casting, hammering into plates, cutting and drawing it into wire, that were in England ... There was likewise a fulling-mill upon the same stream, now demolished; but the hammer for iron remains. These I mention because I do not remember to have seen such a variety of mills and works upon so narrow a brook, and in so little compass' [Fig.3.4].[6] Richard Evelyn did not continue the gunpowder manufacture with which his family had been associated, but he did acquire an iron forge and set up a copper-mill and wire works, the latter on the site of a large pond adjacent to Wotton House called the pigeon-house pond, also shown in one of Evelyn's drawings [Fig.3.5].[7]

FIG.3.4: EXTRACT FROM THE MAP IN JOHN AUBREY, *NATURAL HISTORY AND ANTIQUITIES OF THE COUNTY OF SURREY* (1819) SHOWING THE MILLS ALONG THE RIVER TILLINGBOURNE

FIG.3.5: 'THE PROSPECT OF THE OLD HOUSE AT WOTTON 1640 FROM THE BROOME FIELD'. DRAWING BY JOHN EVELYN SHOWING THE PIGEON HOUSE AND LARGE POND.

But to Evelyn and his brother, hills, wood and water suggested quite different possibilities: namely 'all the amoenities of a villa and garden after the Italian manner', which would make Wotton comparable to the best country seats in the nation.[8] From the time John was a schoolboy he had what he calls an 'extrordinary fansy to drawing and designing', though his father discouraged it as a distraction from his serious studies.[9] To avoid involvement in the early years of the Civil War, he spent some time travelling in the Low Countries. Returning in 1643 he took refuge at Wotton and, while he waited for an outcome of the struggle between King and Parliament, occupied himself with making what he describes as a 'triangular Pond or little stew, with an artificial rock', and a little study or garden house 'over a Cascade, to passe my Malencholy houres shaded there with Trees'. A crayon drawing survives of this charming, fragile-looking structure, looking like a little tree-house reached by a spiral staircase. He had probably got the idea from something he had seen in the great gardens of the Low Countries, and it was, as he says, a trifle. But 'this trifle', he adds, '... was the Occasion of my Bro: vast Expence, when some yeares after, he Inlarged the Gardens, built the Portico, & Cutt the Mount into the present shape it now is of, with the fountaines in the Parterr, which were amenitys not frequent in the best Noble mens Gardens in England'.[10]

By this time John had gone to the Continent again, to spend the next several years in France and Italy. George's designs he says, were 'finished whilst I was abroad ... conducted by a Relation of ours, George Evelyn, who had ben in Italy, but was mistaken in the Architecture of the Portico, which tho' making a magnificent shew, has great faults in the Colonade, both as to the Order, which should have ben Corinthian, & the Ornaments, the rest is very tolerable'. This creation involved moving what Evelyn described as a 'great mound overgrown with trees close to the house' (clearly shown in his first drawing of Wotton), by digging away the earth and re-distributing it by means of a fast-flowing adjacent stream. Thus the old moat was filled up and a level area created for a parterre before the south front of the house. A leat was then drawn off the Tillingbourne at a point 50 feet higher than the parterre, which supplied the fountain at its centre by gravity. The remaining raised ground to the south of the house was cut into terraces and the face towards the house supplied with a grotto, inset with statues in niches and fronted by the massive stone portico whose classical correctness Evelyn criticized [Figs.3.6, 7 & 8]. Even so, he claimed that they became 'the most famous of England at that tyme'[11] and, of course, they are the features that remain substantially with us today.

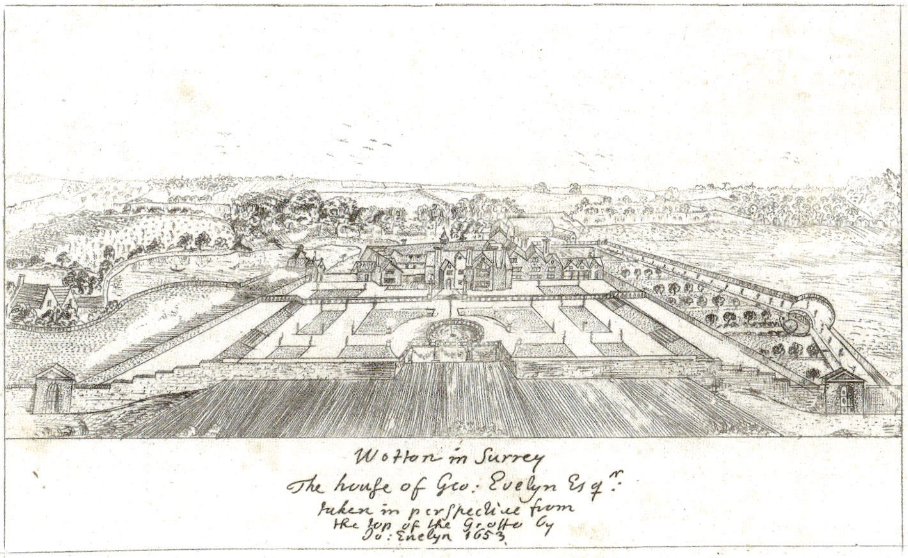

FIG.3.6: ETCHING OF A DRAWING BY JOHN EVELYN
OF WOTTON HOUSE AND GARDENS, *CIRCA* 1653

FIG.3.7: WOTTON IN 2006
SHOWING THE TEMPLE AND MOUNT

This is the account in the memoir Evelyn wrote towards the end of his life. However Alastair and Carola Small have pointed out in their article on Wotton as Epicurean garden in the *Journal of the Warburg and Courtauld Institute* that Evelyn flatly contradicts himself in his diary account, by saying that this work, far from being done in his absence, was mostly carried out after his return from abroad in 1652, and under his direction. The entry in question is that for 22 February 1652, in which he says:

> I went with my Bro Evelyn to Wotton to give him what directions I was able about his Garden, which he was now desirous to put into some forme: but for which he was to remove a mountaine that was over-growne with huge trees, & thickett, with a moate, within 10 yards of the very house: This my brother immediately attempted, & that without greate Cost for more than a hundred yards South, by Digging downe the Mountaine, & flinging it into a rapid streame, which not onely carried the Land &c away, but filled up the moate, & leveld that noble area where now the Garden and fountaine is.[12]

FIG.3.8: WOTTON IN 2006 FROM THE MOUNT

The Smalls conclude that the latter account must be the correct one, as it agrees with John Aubrey's account of Wotton in his *Natural History and Antiquities of the County of Surrey*, which was written from Evelyn's information (see

Appendix). They argue that the reason Evelyn later denied his involvement in the garden was that he had by then rejected the Epicurean philosophy with which aspects of the design could be associated and did not want others to make the connection.[13]

In fact neither the diary nor the memoir (nor Aubrey for that matter) are very good evidence, as they were all (including the diary in the form we now have it) written retrospectively. John's letters to his brother for the relevant period have long vanished, along with the rest of George Evelyn's personal archive. But John kept his correspondence and the letters George wrote to him in Paris in 1651, when work on the gardens began, do survive, and from them it becomes much clearer what the chronology of the garden's construction, and John Evelyn's part in it, actually were. The first evidence we have is a letter from George of 12 December 1650. He first asks his brother for some melon [Fig.3.9] and cypress seeds, and then for further information about 'what green tree it is, that you informed me you had frequently in the Gardens in France which grew suddenly into a fine hedge & kept green all the year'. If John can send him plants or seeds of this kind, he adds,

> you will infinitly oblige me: & give great encouragement to the worke I have in hand which is levelling my hill behind the house & makeing a Bowlin green; some Walkes & other devises, All [which?] my Cousen Evelyn directs; so that I hope it wilbe made very [word illegible], & render my seat very advantagious both for present & [futur]e. The bankes of my Walkes I would willingly set with some green Larell or the tree you mentioned, that it may looke green always: I intend to make upon my hill a very spatious Grotto & fountaines below, my Water & ground beeing apt for such things: if you can collect some curious shells or stones or pebbles: it will be very serviceable to me:

FIG.3.9: MELONS BY JOHANN WILHELM VON WEINMANN FROM *PHYTANTHOZA ICONOGRAPHIA*, 1735

> in regard I mean to make it very Naturall: I will have the Water to come into it: & all along having my Cousen['s] advise I hope I shall doe nothing ill, a fault which formerly I have ben to guilty of.[14]

By this time John had been abroad for several years and, although he made one fairly prolonged visit to England after his marriage to the daughter of the English Resident in Paris in 1647 and although he records cousin George visiting him in Paris in 1649, the explanatory tone of brother George's letter does not suggest that this project was something he or cousin George had consulted John about previously. At this point it is the detail, the evergreen planting and the lining of the grotto, matters in which John had taken a particular interest on his travels, that George is consulting him about.

We don't have John's reply, although he obviously regarded it as important, since, unusually for him, he recorded writing the letter in his diary. But he had made copious diary notes of garden grottoes he had visited in France and Italy, and from these and the details he recorded in his later unfinished 'History of Trades' manuscript, we know what he is likely to have recommended. In the latter, under the heading 'Grotesco' (which he defines as the 'Grotto or shell worke, wherewith Fountaynes &c & all rocke worke is made'), he writes that 'the mayne thing is to doe it most natural, such as is to be seene in divers places beyond the sea, as with all sorts of sea shells, branches of Corral reall or artificiall, Marcasits, glases, mother of pearle, mosse & Ivy Naturall or Artificial'. The ground work, he adds, should be of brick or flint, to which the decorative materials are fastened with a cement, for which he gives the following recipe: 'New yellow way a quarter of a pound, Resin one pound, Pitch of Burgogne half & quarter of a pound, Poudr of tyle extreamly small beaten & refined, as much as suffices to make the former to a working Consistence'.[15]

George's response indicates that John wrote something like this in his letter, together with much detail of examples he had seen on his travels:

> Sir, I must return you my harty thankes for your great perpensity to furnish me with Materialls for my intended Grotta: as also for your noble advise: which I shall endeavor in all particulars to observe. Sir, you have furnished me with such instructions in this affaire; that it bespeakes you a great observer of all things; & a most exact Travailer: I wish our Country afforded me with all such gallant materialls as your letters dictate: you may be

confident, I shall be very solicitous to procure all our parts can afford mee & what my freinds can helpe me too: both there & in other parts: I have desired a Barbadoes Merchant to furnish me with the curious Rocks of that Iland & my Corall, Conchas & other shells: what our ... shoours affords I will spare no cost to obtaine: & what my owne or my freinds ground affords. You have very rightly observed the more natural or wilde it is invented: the best, & most proper for such a designe: & for naturall greens as Ivy & the like, I shalbe very diligent to see it inserted in the worke.

John had evidently then gone on to advise in detail about the water-works, for George adds:

for the water I shall punctually observe your directions: I have the comand so of my smal stream, that I hope shal compasse all those varietys, with litile cost, & to the most exact advantage. Sir, this is a worke of some cost, & no litile tyme it will take up to furnish my selfe with Materialls proper for it: therefor I shall proceed with it very leisurly: it will not be a worke of this yeare. I intend to cutt out the plott now: & if it can be effected to Wall the sids & arch it this sumer if not the next: but to finish it will take up a longer tyme: so that I hope I shall have the happines to have your personal advise & direction. I shall make it a very spacious one: & every way compleat. You know my Architector: I hope I shall not err: as I can gett materialls so I will proceed. my Cousen shall send you the plott. I hope it will please you ... I have a Draught of it, drawen by my Cousen.

The evergreen George wished to introduce into his garden on his brother's recommendation is at this point identified as the alaternus (or Italian buckthorn [Fig.3.1]), which John had seen in France and was passionately fond of: the best for 'hedges and verdure in the world (the swiftness of the growth consider'd) ... and the honey-breathing blossoms ... an early and marvellous relief to the bees'.[16]

[I shall be?] glad to rec. from you the seeds of the Green Alaternys [George continues] which I will plant [ ] to cover all my banks: you having had so full experience of its verdure & sweetnes: I cannot thinke any thing can be more proper for those places, my desire being to have those places render me a continued summer. What I doe now is in order to compleat my seat; that my estate [i.e., income] will make me any way capable of so handsome a designe: I despaire of that & leave it to posterity. [If your] acquaintance or interest can procure me any curious shells or other Materials proper for this worke I shalbe very [thankful] for them: & returne you the moneys in the purchase of them.[17]

John obviously responded by offering to procure a further evergreen species, and with details of where to obtain the grotto materials, for in February 1651 George replies:

> I shall make a diligent search after the Phylaria: & if they are to be procured I shall not put you to the trouble ... I would be glad to plant such hedges as would alwayes by its greenes give us a spring ... as to the Materialls for the grotto: now I am enformed by you where they are to be procured. I shall when I have built it empl[oy] some freind in those parts to collect them for me & when I [come?] to finish it, I shall take my directions from your better genius. This wilbe the worke of another year: & their is no hast in doing [it], but my resolutions are, when I doe build it, to do artificially [both?] for the water workes, & shell worke.[18]

Then there is a gap in the gardening letters, until in September of the same year George writes:

> though I have not troubled you with paper, yet I have you dayly in my conversation: & not a cup passeth but your friends remember you ... I am perfecting my designe ... the only thing that is wanting will be moneys for [it is?] capable of many improvements had I purse to answer it ... My first designe was a bowling green, but the Ladys vote will carry it for a garden, & in it a fountayne: haveing the advantage of water. But I shall not proceed therein without your approbation & advise; with directions when I have sent the plott, how to designe my fountaine [and?] raritys of waterworkes: I have almost levelled the hill behind the house & built my walls: the next worke wilbe [to] designe the ground for a garden & fountayne: The plott I am resolved to send you: & then begg your advise and observations.[19]

In December 1651 he reports that an aviary has been added to the design, 'to be built with small charge', but that the alaternus seed John had sent came short of expectation through his gardener's lack of expertise. The promised plan of the garden is also delayed, because cousin George, having not yet quite finished it, 'was unwilling to present you with imperfections'. But by this time John's return to England was imminent and George hopes he will make this unnecessary by coming to Wotton in person. Finally he adds, 'what seeds you can furnish us at spring of the year to make perpetual verdures or what choice flowers you can procure [I] shalbe happy in their production'.[20]

What we have to conclude from these letters is that the overall grand design, essentially an engineering and architectural one, of levelling the hill on the south side, filling the moat and creating the area for the parterre and fountain, were conceived and largely carried out in John's absence and under Captain George Evelyn's direction, with little input from John [Fig.3.10]. It was chiefly the details of planting and the ornamentation of the grotto and fountain with which George sought his help. However we have to note the absence in George's letters of any mention of the massive portico fronting the grotto; the only structures he

FIG.3.10: ENGRAVING OF WOTTON HOUSE FROM *A TOPOGRAPHICAL HISTORY OF SURREY* BY E.W. BRAYLEY, ED. EDWARD WALFORD (1848)

mentions are walls, which must refer to those dividing the mount from the level parterre. It is also clear that at this stage he conceived the grotto as lined with shell-work; there is no mention of the niches and statuary. As the portico can only have been created after the shaping and walling of the mound had been completed and this was still in progress in 1651, this suggests that it is to be dated after Evelyn's return. Alastair and Carola Small point to a description of a grotto with statues in John Evelyn's great unfinished gardening work, 'Elysium Britannicum', which sounds very similar to that at Wotton, and cite this as evidence that the latter was John's 'overall conception'.[21] However the only surviving manuscript of the 'Elysium Britannicum' dates from the early 1660s, several years after the Wotton design was complete, and so this reference may simply be a description of it. Against the idea must be set John's clear statement that it was Captain George's work and his blunt criticism of his cousin for mixing the classical orders and general 'over-building'. If it had really been John's conception, we have to suppose that he would have been able to correct these matters at source.

This design apparently meant that the naturalistic shell-work lining for the grotto, discussed in the brothers' letters, was abandoned, but Evelyn did probably take a considerable part in ornamenting the parterre fountain and in the planting. The latter is confirmed by a letter from George three years later, anticipating another visit from John, this time with his wife and son:

> there is not a Flower in our Garden but you may command: & for the Greenes we have already sett out offsetts for you: & for the Gillyflowers [Fig.3.11] you shall have slips of each sort when tis serviceable tyme to take them: and I trust you wilbe so kind to us as to come and bring my sister & pretty Dicke & spend some part of this summer with us: and at parting wee will have all the garden lookt over to serve you: The best way wilbe to come & make your own choise: for it cannot be better bestowed then on you, to whom I am very much indebted upon the same score.[22]

FIG.3.11: CARNATIONS (GILLYFLOWERS) BY CLAUDE AUBRIET (1665-1742).

In *Directions for the Gardiner at Says-Court* Evelyn wrote :
'To have stock Gillyflowers very faire, choose a plant which beares excellent double flowers, suffering it to beare but only one branch of flowers: save and sow the seedes of those flowers in February on the Hot-bed, and plant them forth in Michaelmas: This is a precious seacret.'

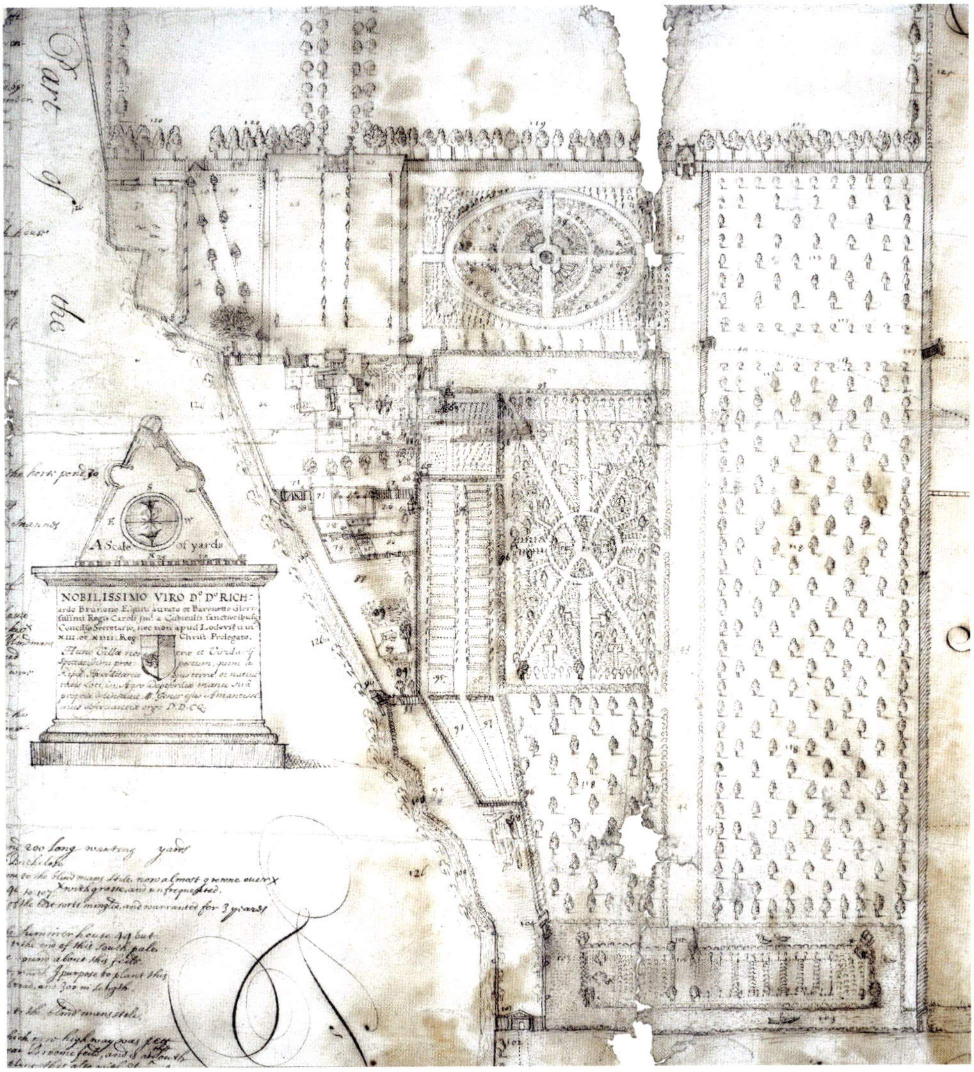

FIG.3.12: PLAN OF SAYES COURT, DEPTFORD, *CIRCA* 1653

As this indicates, John had gone on by this time to establish his own garden in the most unpromising of sites: at Sayes Court next to the dockyard at Deptford [Fig.3.12]. He did not undertake any of the structural works there which made Wotton so striking. What Sayes Court most conspicuously lacked, given the flat and featureless terrain, were grottoes and high places. To try to create these artificially would have meant expense far beyond John's means, and indeed his expertise; as a gardener he was more notable as a plantsman than an architect or engineer. But in a description of a garden otherwise very like Sayes Court (even to its garden laboratory) which he added to his translation of a little French alchemical work, he gave it some imaginary grottoes and even a 'stupendious precipice' with a natural water supply, perhaps transporting them in his imagination from Wotton.[23]

FIG.3.13: CYPRESS FROM *DELICIAE ET ARCANA FLORUM* BY ANDREAS DE LA CROYX (MAGDEBURG, 1710)

Otherwise Sayes Court was essentially a garden of trees, and especially of evergreens. From his travels in Mediterranean countries John had brought home with him a longing for a garden which did not die or become dormant for half the year. In a northern climate the cultivation of evergreens was of course one means by which art could supplement nature. They protected the house with their foliage all year round and produced an Italianate illusion (at least when looked at from indoors) of perpetually mild climate, 'for which reguard', Evelyn added, 'we pronounce it a most sweete & incomparable part of the Hortulane amoenitie and to our particular inclination, the most agreeable'.[24] None of the deciduous natives, however stately, were comparable for him to those 'which are perpetually greene', the laurel, cypress [Fig.3.13], fir, yew [Fig.3.15], holly [Fig.3.14], as well as the more delicate alaternus [Fig.3.1] and phylaria which he had recommended to his brother and helped to naturalize in England.

In fact probably the most important connection between John Evelyn and Wotton was not his influence on the garden, but the influence of the whole topography on him, chiefly in fostering this passion for trees. He describes himself as 'wood-born', and dates his love of wooded landscape from his earliest years at Wotton. A deeply religious man and ambivalent about human desires and appetites, Evelyn always found the inexorable strength and self sufficiency of the tree immensely appealing. He never tired of celebrating the mystery by which the tiny grain, 'which lately a single Ant would easily have born to his little Cavern', with its 'insensible rudiment, or rather habituous spirit', would ascend 'by little and little ... into an hard erect stem of comely dimensions, into a solid Tower as it were'. And then, he goes on, 'consider how it assimilates, separates and distributes ... how it concocts, transmutes, augments produces and nourishes without separation of Excrements (at least to us visible) and generates its like without violation of Virginity'. Thus the tree could very readily become the embodiment of the higher nature and the spiritual aspiration of man. If singly trees were always 'sacred Shades' to Evelyn, groves were nature's temples, acknowledged in Christian and classical antiquity as numinous, predisposing to reverence, meditation and prayer. After the 1650s he extended his tree-planting at Sayes Court, so that the long promenade which originally bounded the grove became the central walk of an extended plantation in which the evergreens predominated.[25] The same passion led to his *Sylva: or a Discourse of Forest-Trees*, in its four ever-enlarging editions. This was the first publication to be sponsored by the Royal Society and is still Evelyn's most attractive and impor-

FIG.3.14: *ILEX AQUIFOLIUM* (HOLLY) BY PIETRO ANDREA GREGORIO MATTHIOLI (1501-77)

tant work. Intended to encourage large-scale tree-planting both as an economic resource and an amenity, it appears to have had a remarkable influence on landowners in his own day and later.

While Evelyn consorted with Kings and ministers, became a founder member of the Royal Society and a notable public servant after the Restoration, his brother George remained at Wotton, marrying twice and begetting several children. But all did not go well with him. Both his wives predeceased him, leaving him a widower while he was only in middle life. Both his sons were ne'er-do-wells and his one surviving daughter Mary was not much more comfort. Much less evidence than one would expect survives of his fifty-eight year tenure at Wotton, but the most evocative are the letters he wrote her whenever he was away from home. Though they have nothing to do with the gardens, I can't resist quoting from them. It would have been natural for Mary to have succeeded to her mother's house-keeping role as soon as she was old enough. Yet it is clear from the increasingly exasperated sequence that she was a strong-charactered and rather eccentric young woman who had no intention of being burdened with this task, either by her father or by any husband her family might have in mind for her. At first her father's tone was mild and coaxing:

> I would have you this fine weather rise early & keepe good houres & come down in the morning amongst the maids that they may see that you mind good housewifery: ... I would have you ... set not up too late at nights & serve God in the first place. ... Your company now being gone, you must now contract the expences of the house, and have but a joint of mutton at night, there beeing none but you and your sister ... I would have you goe to bed at a good hour & rise now at seaven in the morning, that soe you may walke & take the fresh aire which will infallibly take away the pimples in your face ... Goe to bed in good tyme, & if you give that example, all will follow your way, twill save candle and fire.[26]

Later the subject matter is essentially the same, but the tone is more severe:

> I should thinke my selfe very happy, if you would so far obey me as to breake your selfe of that ill custome of going late to Bed & rising late in the morning ... tis a great shame to you & if you knew how much it troubles me these bad houres, you would certainly be so obedient & discreete as to rise as other young Ladyes do.[27]

FIG.3.15: THE YEW TREE FROM *SILVA*, 1801 EDITION

Evelyn in *Sylva* (1679 edition) writes of the Yew: 'He that in winter should behold some of our highest Hills in Surrey, clad with whole Woods of these … Trees, for divers Miles in circuit (as in those delicious Groves of them, belonging to the Honourable, my noble Friend Sir Adam Brown of Bech-worth-Castle, from Box-hill, and neer our famous Mole or Swallow) might without the least violence to his Imagination, easily phansie himself transported into some new, or enchanted Country.'

And finally, in extreme exasperation:

> Let me now give you my commands, which are that you serve God and that every Sunday you rise so early as to be at the church to hear the common prayer and that your mend your houres of rising and come to dinner by twelve a clock. If you do not observe my commands in this, expect no blessing from God & until I hear you do what I desire ... I am resolved never to see you, but leave you to yourself. Tis a lazy habit you have got to lye a bed. No woman in the nation doth it but yourself and therefore quite it except you will be a scandal to yourself and a griefe to your father.[28]

The wife of the parish clergyman had to act as housekeeper at Wotton instead. When she died, George Evelyn offered his daughter the profits of dripping and tallow from the household as a personal income, to encourage her, as he put it, 'to look after things' herself from now on. But to no avail. Various eligible offers of marriage were made, but her father never forced her choice, 'that hereafter you may not say, to please your freinds you marryed a man you could not love'.[29] And so she was able to evade them. She did finally marry, but not till late in life and to a husband who was often abroad on government service, so that he disturbed her comfortable way of life at Wotton as little as possible.

As George Evelyn's rather ineffectual preoccupation with saving fire and candle and expenditure on meat suggests, all was by this time not well at Wotton. His brother says that the creation of the garden had been a great expense to him, and while this would certainly have been true, it was not the only factor. In fact the landscape which had inspired John as a child had gradually been transformed from continuous woodland into a series of coppices planted for the use of the local industry. Standard trees of course will not regenerate from coppices, and in large areas the landscape simply degenerated into heathlands, which then also came under constant pressure for fuel. In his *Sylva* Evelyn described these as 'cold starving places, neglected and despaired ... hopeless grounds, barren hills as any in England that lately have borne woods, groves and coppices'.[30] By the end of the century most of the remaining trees had been felled to pay debts, and Wotton was left, in John's words, 'naked and ashamed almost to own its name'.[31]

Since George Evelyn's two sons predeceased him, John in old age found himself heir to Wotton by right of entail. His own finances had by this time been greatly reduced by a prolonged law-suit involving the Sayes

Court estate, and he proposed to George that he and his wife should come to join households with him at Wotton, 'in the Place that gave me a Birth, & must shortly Burial'. The famous garden at Sayes Court was by this time in decline, but John, in his seventies, was still prepared for a new horticultural project. He begged George 'to gratifie my Inclynations by Devolving on me the Care, & Supervisal of your Gardens, and the Culture & Adornment of the Place of our worthy Fore-fathers: The mutual Enjoyment of those Innocent and primitive Refreshments (with Books, and such a Friend) being all my great Age is Capable or indeed Ambitious of'. In advance he sent some seedling trees for planting, with directions from his gardener; 'together with a Draught of a plot representing my thoughts in order to a Grove of Evergreenes ... To this Viretum (the only Desiderate I think to Accomplish your Gardens at Wotton) I would also add a Conservatorie for Orange, Limon & other rare & choice shrubs.'[32]

When he finally came to own Wotton outright at the end of century, he pledged himself, old as he was, to begin repairing the damage to the whole estate: 'I shall if God protract my years and continue health be continually planting, till it please him to transplant me to the glorious regions above, the celestial Paradise, planted with perennial groves, and trees bearing immortal fruit'.[33] Though he had very few years at Wotton himself, he left a like-minded heir, his grandson John, brought up by him and moulded by his precepts. In a little advice manual Evelyn wrote for him just before he died, he exhorted the young man to preserve the gardens and above all plant trees, 'which in a few years will prove of incredible Emolument & restore the Name of Wotton, otherwise in danger to be lost and forgotten, planting of Timber-trees being in truth the onely best and proper Husbandry the Estate is capable of'.[34] The youngest John Evelyn took all this to heart. He succeeded at Wotton in his twenties in 1706 and lived there until 1763. His archive, like that of his great-grandfather Richard, indicates that his overriding concern was the improvement of his estate. Apart from the family letters, it consists of an almost unbroken sequence of correspondence with his estate stewards. He made renovations to the house, including a new library. Without disturbing the centre-piece of the Italianate garden, he introduced a more naturalistic landscape involving water, serpentining the Tillingbourne brooks and ornamenting them with rock-work,[35] and his monument in the church pays tribute to his planting. His own grandson, Frederick, 3rd Bart., died without male heirs and the baronetcy passed successively to two of his cousins, brothers who were both of unsound mind. So the estate was deliberately left to a different branch of the family, distant cousins, and by the mid-nineteenth century one of them, William John Evelyn was in possession. His was another very long tenure at Wotton, lasting until the early years of the twentieth century, and it was he who must have been responsible for the garden restoration which is still apparent, adding the Pulhamite stone and so preserving the essentials of the seventeenth-century design until today [Fig 3.16].

FIG.3.16: PART OF THE PULHAMITE ROCKWORK AT WOTTON PHOTOGRAPHED IN 2005

John Evelyn wrote much about garden philosophy, garden theory and garden practice. Sayes Court has long vanished and it is natural to want to attribute to him an important surviving garden. On the other hand we know frustratingly little about cousin George Evelyn; not even for certain what his place in the family was.[36] It is only from John's passing reference that we even know he travelled in Italy. Yet the evidence suggests that it was this cousin, not John, who was largely responsible for the remarkable Italianate design which is Wotton's most notable surviving feature today.

REFERENCES

1 British Library Add. MS 78431: John to Mary Evelyn, 1 May 1660.

2 *The Miscellaneous Works of John Evelyn*, ed. William Upcott (London, 1825), p. 688: Evelyn to John Aubrey, 8 February 1676.

3 E.S. de Beer (ed.), *The Diary of John Evelyn*, 6 vols. (Oxford, 1955), vol. 2, p. 2.

4 Ibid., vol. 1, p. 5, vol. 2, p. 4.

5 BL Add. MS 78610 A.

6 *Miscellaneous Works*, ed. Upcott, pp. 689-90: letter to Aubrey, 8 February 1676.

7 BL Add. MS 78610 B.

8 *Miscellaneous Works*, ed. Upcott, p. 687: letter to Aubrey, 8 February 1676.

9 *Diary*, vol. 2, pp. 9, 22; W.G. Hiscock, *John Evelyn and his Family Circle* (London, 1955), p. 7.

10 *Diary*, vol. 1, p. 55; BL Add. MS 78610 C.
11 *Diary*, vol. 2, p. 81.
12 *Diary*, vol. 2, pp. 60-61.
13 Carola and Alastair Small, 'Evelyn and the Garden of Epicurus', *Journal of the Warburg and Courtauld Institute*, vol. 60 (1997), pp. 198-202, 212-4.
14 BL Add. MS 78303, f. 49: George to John Evelyn, 12 December [1650].
15 BL Add. MS 78339, f. 251v.
16 *Sylva* (1706), ed. John Nisbet (1908), vol. 1, p. 282.
17 BL Add. MS 78303, f. 50: George to John Evelyn, 12 January 1650/1.
18 BL Add. MS 78303, f. 52: George to John Evelyn, 6 February 1650/1.
19 BL Add. MS 78303, f. 57: George to John Evelyn, 29 September 1651.
20 BL Add. MS 78303, f. 58: George to John Evelyn, [December?] 1651.
21 Small, 'Evelyn and the Garden of Epicurus', pp. 200-01.
22 BL Add. MS 78303, f. 75: George to John Evelyn, 9 July 1655.
23 BL Add. MS 78346, ff. 3-6: preface to 'Coelum sanitatis'.
24 John Evelyn, *Elysium Britannicum*, ed. John Ingram (Philadelphia, Pennsylvania University Press, 2001), pp. 313-15.
25 F. Harris, *Transformations of Love: the Friendship of John Evelyn and Margaret Godolphin* (Oxford, 2003), pp. 26-30.
26 BL Add. MS 78295, f. 107: George to Mary Evelyn, 24 January 1670.
27 BL Add. MS 78295, f. 113: George to Mary Evelyn, 19 September 1671
28 BL Add. MS 78295, f. 121: George to Mary Evelyn, [*circa* 1671].
29 BL Add. MS 78295, f. 127: George to Mary Evelyn, 20 November 1672.
30 Peter Brandon, 'Land, Technology and Water Management in the Tillingbourne Valley', *Southern History*, vol. 6 (1984), pp. 90-99.
31 Evelyn, *Memoires for my Grand-son*, ed. Geoffrey Keynes (1926), p. 17.
32 BL Add. MS 78291: John to George Evelyn, 21 November 1692.
33 *Sylva* (1706), ed. John Nisbet (1908), vol. 2, p. 260.
34 *Memoires for my Grand-son*, p. 17.
35 Brandon, 'Land, Technology and Water Management ...', p. 90.
36 G.D. Scull, *The Evelyns in America* (1881), pp. 1-45, 355-357, identifies him as the son of John Evelyn's uncle, Robert Evelyn of Godstone; Esmond de Beer suggests that he was the son of another uncle, Thomas Evelyn of Long Ditton; see Evelyn, *Diary*, vol. 2, p. 551, note 3.

FRANCES HARRIS is Head of Modern Historical Manuscripts at the British Library and has been curator of the Evelyn Papers since their acquisition by the British Library in 1995. She is editor (with Michael Hunter) of *John Evelyn and his Milieu* (British Library, 2003). Her publications include 'The Manuscripts of John Evelyn's *Elysium Britannicum*', *Garden History*, vol. 25, no. 2 (1997), and *Transformations of Love: the Friendship of John Evelyn and Margaret Godolphin* (Oxford University Press, 2004).

FIG.4.1: A CHIMPANZEE FROM *ORANG-OUTANG, SIVE HOMO SYLVESTRIS: OR, THE ANATOMY OF A PYGMIE COMPARED WITH THAT OF A MONKEY, AN APE, AND A MAN* BY EDWARD TYSON (1699)

4

'WHOLLY NEW AND AMBIGUOUS': THE DISCOURSE OF NATURE

DOUGLAS CHAMBERS

John Evelyn was not on what we would now call 'the cutting edge' of science. In the seventeenth century the line between gardening and science was no easier to discriminate than the line between science and the marvellous. Early in his famous *Diary*, for Rome in 1645, Evelyn writes of a mountebank (showman) in the Piazza Navona who took from his finger a ring, wetted it, and touched it into flame.[1] But had not Galileo thirty years earlier concluded his account of his momentous discoveries with the telescope by taking his friends into a half-darkened room and showing them a similar marvel using what we now know as barium sulfide?[2] [Fig.4.2] If the categories of science were elusive, what were its boundaries?

FIG.4.2: GALILEO WITH HIS TELESCOPE IN THE PIAZZA SAN MARCO, VENICE
WOOD ENGRAVING

Galileo's interest in the transmission of light had been shared by his friends and defenders, Federico Cesi and Francesco Stelluti. But these men were even more interested in the origins of life: what we would now call evolution. The ambiguous nature of fossilized wood that interested them (and was previously thought inexplicable), threatened to erode the old certainties of Aristotelian classification let alone the church's teaching on biblical history. If these boundaries were permeable, if one substance could change into another, nature itself was mutable and could no longer be thought to be exempt from change. With that insight came the realization that the earth itself was not static and the increasing conviction that the planet had evolved.

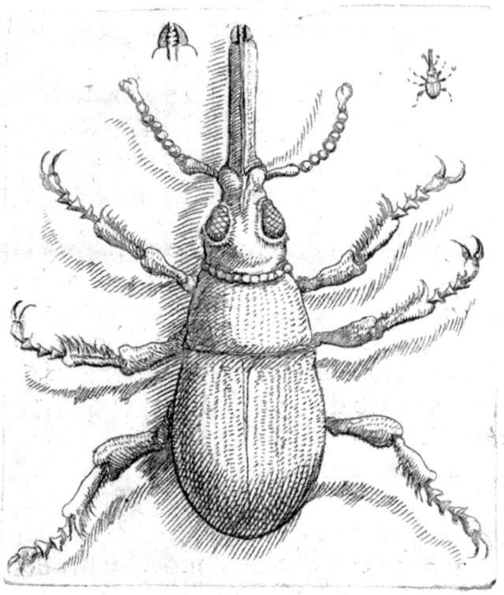

FIG.4.3: BEETLE FROM *PERSIO TRADOTTO IN VERSO SCIOLTO E DICHIARATO* BY FRANCESCO STELLUTI (1630)

Into the wake of this 'evolutionary moment' the young John Evelyn was ushered on 21 November 1644[3] when he visited the apartments of the great Roman collector and scholar, Cassiano dal Pozzo. Among the 'divers folios' he was probably shown there was Cassiano's collection of natural history drawings [Fig.4.3] that were eventually (in 1762) to become part of the collection of George III. Unlike most of the 24-year-old Evelyn's entries in his early *Diary*, however, his account of this day was not taken subsequently from a published guidebook (*Diary*, vol. 2, p. 277, n. 3). He wrote of what he saw. And certainly the drawings made an impact on him, for he recognized them and their origin in Cassiano's collection when, under the auspices of the London physician Sir George Ent (1604-89), they later appeared in London.

What Evelyn stepped into in 1644 (though probably he did not recognize it at the time) was not only the salon of a great connoisseur of the arts but the scientific *studiolo* of one of the strong supporters of research into fossils – in this case the fossils of trees. Cassiano's interest in these fossils was one part of the late Renaissance argument about evolution. Not only was Cassiano a defender of his old friend Federico Cesi against the denunciations of the Inquisition, but he also endorsed the ambiguous nature of Cesi's discoveries. Were these fossils wood or stone, and how did this fossilized stone *(legno minerale)* challenge previous fixed beliefs about nature and the origins of Creation itself: an account in which evolution had had no place.[4]

Cassiano's scientific interests came from his still potent memory of Galileo's earlier fossil researches and the work of Galileo's disciple, Francesco Stelluti on fossilized minerals: *Trattato del Legno Fossile Minerale* (1637). This work was dedicated to Francesco Barberini, another patron of the arts and sciences, whom John Milton had also just met in Rome. Stelluti described this fossil research as both 'wholly new and ambiguous', and he wrote about 'the mutable nature of this wood', that is its propensity to change into stone. In fact Stelluti had got his analysis wrong, probably deliberately. His claim that clay had turned into wood instead of the reverse was probably to avoid trouble with the Inquisition. But in fact, as Freedberg has pointed out, Cesi's account nonetheless stood at odds with the traditional biblical accounts both of the Creation and the Deluge.[5]

It was also a challenge to the standard Aristotelian account of nature. If nature was mutable over time, in other words, evolution was its corollary. This recognition of continuous change was a direct challenge both to the church and to Aristotle. Whether Evelyn saw the four volumes of drawings on this subject owned by Cassiano or not, the circle in which he found himself in Rome was one in which the study of fossils [Fig.4.4] was of the highest interest: a study that increasingly recognized the ambiguous categories of description and the process of change in nature.

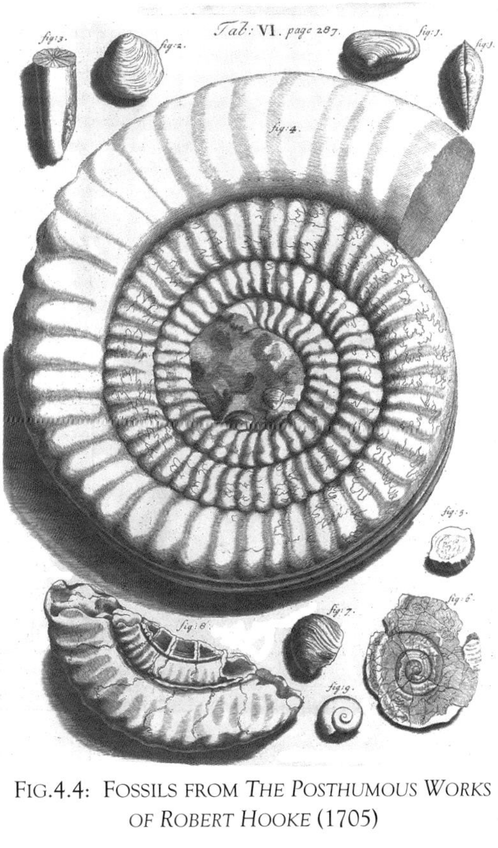

FIG.4.4: FOSSILS FROM *THE POSTHUMOUS WORKS OF ROBERT HOOKE* (1705)

I do not want to exaggerate this moment in Evelyn's life and culture, but it seems to me that it was as momentous as the nineteenth-century moment commonly associated with Lyell and Darwin. Evolution's original meaning in English had the seventeenth-century sense of unrolling a scroll, – as happens in *Paradise Lost* when the angel Raphael gives an account of the Creation. Evelyn was not a theologian, but his theological interests coincided with his encounter with the concept of material evolution: 'the process of developing from a rudimentary to a mature or complete state'. Indeed long before Lyell and Darwin, the word 'evolution' first appeared in the fifth volume of *The Philosophical Transactions of the Royal Society* in 1670.

Nowhere in Evelyn's surviving library, however, is this challenge to traditional certainties of classification more forcefully demonstrated than in Edward Tyson's *Orang-Outang, sive Homo Sylvestris: Or, The Anatomy of a Pygmie Compared with that of a Monkey, an Ape and a Man,* published in 1699.[6] [Fig.4.1] Given to Evelyn by the author (a physician at the two hospitals Bedlam and Bridewell, and a Reader in anatomy at Surgeons Hall), it is inscribed at the end by Evelyn *perlegi* (that is 'read through'). The book's first challenge to traditional classification appears in the dedication to John Lord Somers, then president of the Royal Society. In that society's *Philosophical Transactions* most of Tyson's papers were published. They, too, seem Darwinian almost two centuries before their time.

> 'Tis a true remark [Tyson writes], which we cannot make without Admiration [wonder] that from Minerals to Plants; from Plants, to Animals; and from Animals, to Men; the Transition is so gradual, that there appears a very great Similitude, as well between the meanest Plant and some Minerals, as between the lowest Rank of Men, and the highest of Animals. The Animal [Orang-Outang] of which I have given the Anatomy, coming neerest to Mankind seems to be Nexus of the Animal and Rational.[7] (Evelyn has pencil-marked the last sentence in the margin.)

This is a taxonomial problem that Tyson also found in the writings of the ancients. Philostratus, for example, thought that these creatures could only be '*wild Beasts* of those Countreys', but his predecessor Ctesias 'stiles them Men'. Why has not modern science inquired into these matters, Tyson wonders, in a passage that Evelyn has also pencil-marked. And Tyson goes on:

'Tis an Enquiry that would recompense their Curiosity with abundance of Satisfaction, by the many and useful Discoveries they would make and extreamly enrich the *Natural History* of *Animals*, whose enlargement, I think, in this *Inquisitive Age*, hath not advanced so much as that of *Botanie*. For how great Diligence hath been used of late, to ransack both the *Indies*, to pry into all the Corners of the WORLD, ... to find out a new *Plant*, not before described? And with what great Expence, and how magnificently are their *Figures* [illustrations] Printed? [D3v]

What Tyson offers Evelyn is 'otherness': a challenge to the merely taxonomic impulse to name and categorize the still-unknown natural world. What was troubling about this 'pigmy', the orang-outang, was that he behaved like a man, became used to wearing clothes, and would 'lie in a bed, place his Head on the pillow, and pull Cloaths over him, as a man would do' (p. 8). How is it, Tyson wonders, that the study of such creatures has not advanced in the way that botany has: 'How little hath been done in the Improvement of the *History of Animals*' (p. 2). Here is a creature whose '*whole outward Action is so Humane, and the Passions so lively and significant, that dumb Men can scarce express their Conceptions and Appetites*' (p. 24). As a creature with men 'it would readily eat any thing that was brought to the Table; and very orderly bring it's Plate thither' (p. 30). 'By comparing the *Brain* of our *Pygmie* with that of a Man [Tyson writes] ... it was very surprising to me to find so great a resemblance of the one to the other, that nothing could be more' (p. 54). [Fig.4.5] What troubles Tyson is that the Orang Outang 'our *Pygmie* is no *Man*, not yet the *Common Ape*; but a sort of *Animal* between both': what Alexander Pope was to write of mankind itself – 'a being rudely large

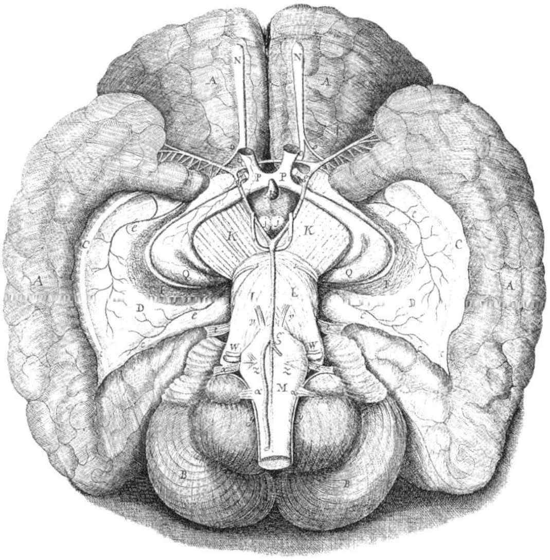

FIG.4.5: HUMAN BRAIN, WITH ITS BASIS UPWARD, AND DIVESTED OF THE DURA AND PIA MATER. LAID OPEN BY DR EDWARD TYSON IN THE COLLEGE OF PHYSICIANS, LONDON. FROM *A SYSTEME OF ANATOMY* (1685)

and darkly great'. All of these passages Evelyn has marked in his copy with his pencil.

Evelyn was well aware that there was much so-called scientific lore that had no more truth than the fabulous mediaeval *Travels of Sir John Mandeville* (1480), but he was plainly fascinated by what was marginal and previously undefined, of which dwarves were one example. In the margins of Tyson's text he writes a puzzled note about a similar English phenomenon:

> So that these two pretty Dwarfs *Gibson* who were married to the dwarf virgins waiting on the Countess of Pembrock ... had Several Children of the stature of other people: and a daughter living mother of a beautiful maid. (p. 29)

This fascination with human size and transmutation is like the later desire of Jonathan Swift's Gulliver to pin nature down. But Tyson does not share Gulliver's certainties about the world. 'I am apt to think', he says, in a passage also marked by Evelyn, 'that *Pan, Aegipan, Silvanus* and *Silenus,* were all the same; as were the *Satyri* and *Fauni,* only these were younger than the former, and the *Nymphs* were the *Females* of the Kind.'

What Tyson exposes, as the modern era begins to establish itself, is the continuing doubt felt about strict definitions of the zoological (and by association, botanical) worlds. In fact these classifications were already becoming outdated, even denied, by continual revisions in classification, revisions as startling in their time as *DNA* has been in ours. If the Italian botanist Ferrar was mad (or at least eccentric) in thinking that each malformed lemon constituted a different species, what are we to make of the differentiations of colour in Newton's spectrum? Ultimately who is making the rules and why? Why do *men* think that the number of *pistils* in a flower determines its species? These are the sorts of question that arose in the almost octogenarian Evelyn's mind as he read Tyson.

At the dawn of what came to be called The Age of Reason, Tyson put his finger on the *superstition* of classifications. And he did so at just the time that Evelyn (and Hooke before him) paused to consider whether such 'mensural enterprises' did not have as much to do with imperial control as with the advance of knowledge. The world that Hooke, Tyson and Evelyn

FIG. 4.6: FRUITS OF LONDON PLANE (*PLATANUS ACERIFOLIA*) INTRODUCED IN 1663

constructed in their research was still marginal. It was not a world divided between the computer's nothing and one. When Evelyn began his garden at Sayes Court, alaternus (buckthorn) and the London plane tree [Fig.4.6] were still unknown in England, and the conservatory uninvented. There was no 'Society of Gardeners' or even a standard taxonomy for plants. Garden design, such as it was, consisted largely of flowerbeds in figures and shapes, and landscape as *landscape* had yet to be discovered. Soil chemistry (as in Evelyn's book *Terra)* had yet to be

FIG.4.7: WOODCUT FROM *LE VRAY ET METHODIQUE COURS DE LA PHYSIQUE RESOLUTIVE VULGAIREMENT DITE CHYMIE* BY ANNIBAL BARLET (1657). JOHN EVELYN ATTENDED CHEMICAL COURSES GIVEN BY BARLET IN PARIS.

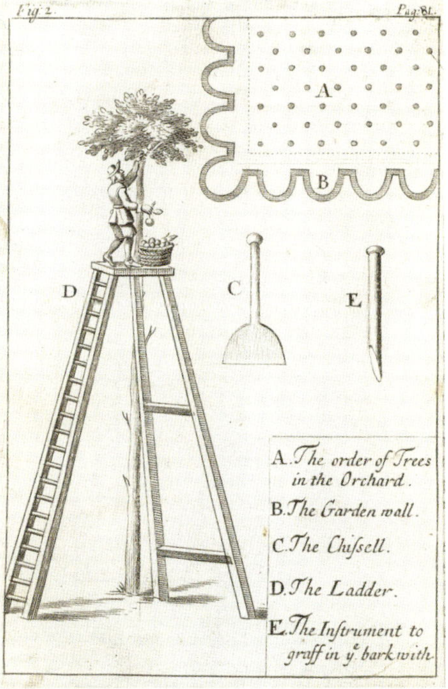

FIG.4.8: ORCHARD MANAGEMENT, THOMAS LANGFORD (1696)

invented, as was the business of fertilising and manuring explained in his *Kalendarium Hortense.* Evelyn's (first English) translation of Lucretius's *De Rerum Natura* introduced the concept of natural change; his *Acetaria* introduced an English readership to the vast range of salad greens [Fig.4.7]; and his *Pomona* greatly expanded the catalogue of fruit available to English orchards [Figs.4.8 & 4.9].

That he never completed his greatest work, on the history of gardens,

FIG.4.9: MALUS *SYLVESTRIS* VAR. *DOMESTICA* BY JOHANN WILHELM VON WEINMANN FROM *PHYTANTHOZA ICONOGRAPHIA* (1735)

'Elysium Britannicum', reflects both the enormous and sudden expansion of botany in his time as well as the impossibility of covering so vast a subject. The book's traces were to be found, however, in at least four of his published books: *Pomona*, *Kalendarium Hortense*, *Terra* and *Acetaria* - books meant to be the foundations of horticultural knowledge and practice ... as was his *Directions for the Gardiner at Says-Court*, not published until 1932 [Fig.4.10]. That there is still much more to be published in the vast archive now in the British Library says more than I can here about his importance to us still.

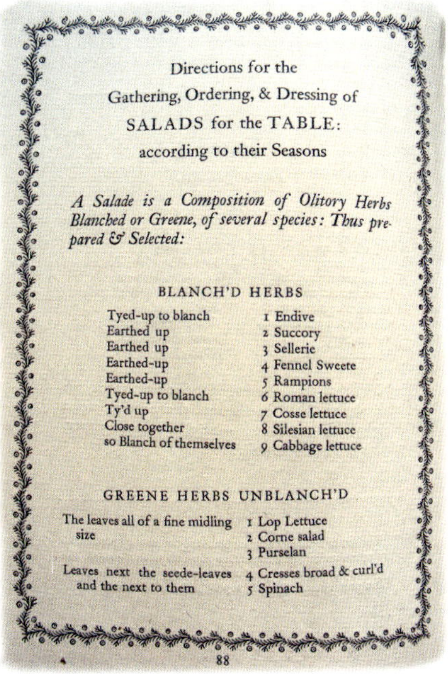

Directions for the
Gathering, Ordering, & Dressing of
SALADS for the TABLE:
according to their Seasons

A Salade is a Composition of Olitory Herbs Blanched or Greene, of several species: Thus prepared & Selected:

BLANCH'D HERBS

Tyed-up to blanch	1 Endive
Earthed up	2 Succory
Earthed up	3 Sellerie
Earthed-up	4 Fennel Sweete
Earthed-up	5 Rampions
Tyed-up to blanch	6 Roman lettuce
Ty'd up	7 Cosse lettuce
Close together	8 Silesian lettuce
so Blanch of themselves	9 Cabbage lettuce

GREENE HERBS UNBLANCH'D

The leaves all of a fine midling size	1 Lop Lettuce 2 Corne salad 3 Purselan
Leaves next the seede-leaves and the next to them	4 Cresses broad & curl'd 5 Spinach

88

FIG.4.10: PAGE FROM *DIRECTIONS FOR THE GARDINER AT SAYS-COURT*

REFERENCES

1 E.S. de Beer (ed.), *The Diary of John Evelyn*, 6 vols. (Oxford, 1955), vol. 2, pp. 397-8.

2 David Freedberg, *The Eye of the Lynx: Galileo, His Friends, and the Beginning of Modern Natural History* (Chicago, University of Chicago Press, 2002), p. 314.

3 *Diary*, vol. 2, pp. 277-8.

4 In fact even Aristotle had recognized that some such process must have occurred.

5 Freedberg, p. 342.

6 The copy in Evelyn's library (Eve.b.17) also contains Tyson's A *Philological Essay Concerning the Pygmies ... of the Ancients.* Evelyn was a friend and frequently dined with him.

7 Dedication to John Lord Somers, f2.

DOUGLAS CHAMBERS is a Professor Emeritus of English at the University of Toronto. His interest in botany is partly inspired by his extensive garden, *Stonyground*, in rural southern Ontario, and he has written extensively on horticultural subjects. He has published a number of articles on garden history, but his best known book is *The Planters of the English Landscape Garden* (Yale, 1993). He has also published an account of his own garden: *Stonyground. The Making of a Canadian Garden* (Knopf, 1996). He is currently editing John Evelyn's Letter Books and is finishing a book on John Evelyn and gardening. This paper, *'Wholly New and Ambiguous': The Discourse of nature*, will form a part of that book.

FIG.5.1: THE CHESTNUT TREE FROM *SILVA* (1776 EDITION)

Evelyn tells us in *Sylva* (1679 edition): 'The Chesnut is (next the Oak) one of the most sought after by the Carpenter and Joyner: It hath formerly built a good part of our ancient Houses in the City of London ... The Chesnut affords the best Stakes, and Poles for Pallisades, Pedaments for Vine-props, and Hops ... I am sure being planted in Hedge-rowes ... or for Avenues to our Countrey-houses, they are a magnificent, and royal Ornament.'

5

A PASSION FOR TREES: THE LEGACY OF JOHN EVELYN

MAGGIE CAMPBELL-CULVER

The title of John Evelyn's 1664 book *Sylva: or a Discourse of Forest-Trees* [Fig.5.2] is in many ways an inaccurate description of what is contained in the book. There is no doubt that the brief from the Royal Society was to focus on trees suitable for converting into industrial timber, in particular for use in ship-building and repair. Timber was then the bed-rock of manufacturing which can be equated to the position that is held by the steady flow of oil today – the least trouble about its supply causes crises. However, Evelyn widened the scope and instructed his readers in the nuances of a great number of trees, only a few of which could be classed as 'forest' or timber trees. The oak, beech, sweet chestnut [Fig.5.1], alder, poplar, the willows, birch and pine, can all be included, both as providing useful timber and as growing in the countryside; but what of the mulberry, walnut [Fig.5.4], cork oak, cedar [Fig.5.17], arbutus [Fig.5.3] and many others written of with enthusiasm by Evelyn?

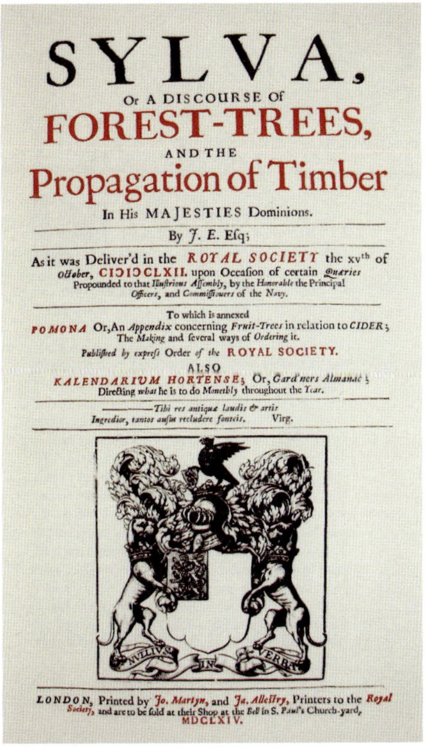
SYLVA,
Or A DISCOURSE Of
FOREST-TREES,
AND THE
Propagation of Timber
In His MAJESTIES Dominions.
By *J. E.* Eſq;
As it was Deliver'd in the *ROYAL SOCIETY* the xv[th] of *October*, CIƆIƆCLXII. upon Occaſion of certain *Quæries* Propounded to that *Illuſtrious Aſſembly*, by the *Honorable* the Principal *Officers*, and *Commiſſioners* of the *Navy*.
To which is annexed
POMONA Or, An *Appendix* concerning *Fruit-Trees* in relation to *CIDER*; The *Making* and ſeveral ways of *Ordering* it.
Publiſhed by expreſs Order *of the* ROYAL SOCIETY.
ALSO
KALENDARIUM HORTENSE; Or, *Gard'ners Almanac*; Directing *what* he is to do *Monethly* throughout the *Year*.
——*Tibi res antiquæ laudis & artis*
Ingredior, tantos auſus recludere fonteis. Virg.

LONDON, Printed by *Jo. Martyn*, and *Ja. Alleſtry*, Printers to the *Royal Society*, and are to be ſold at their Shop at the *Bell* in S. *Paul's* Church-yard, MDCLXIV.

FIG.5.2: TITLE PAGE TO THE FIRST (1664) EDITION OF *SYLVA*

It was Sir Robert Moray (?1608-73), President and one of the founders of the Royal Society, who on 10 September 1662 initiated a discussion regarding the shortage of standing timber and the lack of forward planning to rectify the situation. The following week he had received from four of the 'Honourable Principal

Officers, and Commissioners of the Navy' a paper enumerating their concerns. Sir Robert immediately invited a group of Fellows: Dr Jonathan Goddard (1617-75), Dr Christopher Merret (1614-95), John Winthrop (1606-76) and the youngest, John Evelyn, to consider the problems and find an acceptable solution. It was five weeks later on 15 October 1662 that Evelyn recorded in his journal 'I this day delivered my Discourse concerning Forest-trees to our Society upon the occasion of Certain Queries sent us by the Commissioners of His Majesties Navy'. The lecture made such a powerful impression that the society decided its important message should reach a wider audience, and accordingly Evelyn was 'desired to print the paper read by him'. Surely never in one short sentence has such a seminal book been commissioned.

FIG.5.3 *ARBUTUS UNEDO* BY MARGARET MEEN (1749-1824). WATERCOLOUR ON VELLUM

The resulting work *Sylva: or a Discourse of Forest-Trees, and the Propagation of Timber in His Majesty's Dominions* was published eighteen months later on 16 February 1664: 'I presented my 'Sylva' to the Society; and next day to his Majestie, to whom it was dedicated; also to the Lord Treasurer, and the Lord Chancellor'. Evelyn was both astute and determined to ensure that his message reached not only the ultimate fount of power in the land, the monarch, but also went to the political seats of power and wealth.

The book is a panegyric about the seventeenth-century arboreal world and the core message is, without doubt, how to plant and maintain trees, so that they may play a pivotal supporting role in society [Fig.5.5]. However a dichotomy begins to emerge on the very first page of the Introduction, when he talks of trees that are 'left standing for ornament and service of their country'. What a telling phrase; as throughout the book

FIG.5.4: THE WALNUT TREE FROM *SILVA* (1801 EDITION)

In *Elysium Britannicum*, Evelyn tells us: 'For the most part the Wallnut is an extravagant tree, & suffers nothing to thrive under its shade, ... yet being planted at greate distances if they be choice and straite trees, they become the approches of places {admirably} well, are profitable and of stately aspect.'

FIG.5.5: 'MARCH' FROM *DEN VERSTANDIGHEN HOVENIER* (ANTWERP, 1667)

Evelyn frequently writes of, for instance, 'copses of infinite delight', 'gleams of sun adorned with the distant landscape appearing through the glades', and (discussing the elm) 'this noble tree for shade and delight' [Fig.5.6]. Many of the trees Evelyn discussed, he recommends using for adornment, suggesting the creation of walks, avenues, copses, groves and, a very new concept, vistas. Impressing order on a landscape by planting immense lines of trees was the idea of the moment in England, having earlier come to the fore in Italian gardens, before being further developed in France. Evelyn would have been one of the relatively few people in England at this time who, during his travels on the continent, would have seen these developments *in situ*, and he was full of enthusiasm for such regimental order being imposed. Sweet chestnut '... set them from thirty to ten foot distance', the walnut, which he says 'render most graceful avenues to our country dwellings ...' were to be placed at forty or fifty foot intervals; he enthusiastically recommends the mulberry 'which is very proper for walks and avenues' (presumably except when they drop their

FIG.5.6: THE ELM TREE FROM *SILVA* (1776 EDITION)

purple-juiced fruit). A British native, the service tree, *Sorbus torminalis*, Evelyn believes should be planted as 'the shade is beautiful for walks' [Fig.5.7]. Of the lime tree, he wants to know whether there is 'a more ravishing or delightful object than to behold some entire streets and whole towns planted with these trees'. Many other trees, according to Evelyn, come into the ambit of a planned estate or landscape. He has no inhibitions in welcoming newly introduced trees into either the garden or the

FIG.5.7: THE WILD SERVICE TREE FROM *SILVA* (1801 EDITION)

FIG.5.8: *YUCCA GLORIOSA* FROM PIERRE JOSEPH REDOUTÉ, *LES LILIACÉES* (1802-16)

countryside – in fact he writes of wanting more – so that both pleasure and productivity of the land may increase. He recommends the American *Yucca gloriosa* [Fig.5.8], not for growing as a tree, but to be cultivated as a sturdy hedge, as he does *Punica granatum*, the pomegranate, which he first recorded in his diary as seeing in Paris in 1644 describing it as 'a noble hedge', this native of Afghanistan and Iran is naturalised in south-east Europe, and has been grown in Britain since the fourteenth century. Evelyn promulgated that hedges and hedgerow planting should be treated as not only a safe and secure barrier but also to give pleasure to the eye. Evelyn also has an encompassing eye for including other imported trees including the quince, *Cydonia oblonga* [Fig.5.9], (from south-west Asia, thirteenth century), the strawberry tree, *Arbutus unedo,* from Ireland and southern Europe, and two seventeenth-century arrivals: the tulip tree, *Liriodendron tulipifera*, from North America, and the horse chestnut, *Aesculus hippocastanum,* a native of the borderlands of Greece and Albania [Fig.5.10]. Evelyn also writes of using 'neglected corners of ground' in which to place trees, then a thought-provoking and unusual idea, but now frequently echoed in practice today.

John Evelyn writes discursively throughout the book, although his style to begin with is somewhat oratorical; as he progresses and becomes more confident, the writing becomes more familiar and one can come across surprising and intimate details concerning life in seventeenth-century England: we learn that both beech and sweet chestnut leaves could be used for the stuffing of mattresses, beech leaves 'continue sweet for seven or eight years long; before which time straw becomes musty and

FIG.5.9: THE QUINCE (*CYDONIA OBLONGA*)
PIETRO ANDREA GREGORIO MATTHIOLI (KREUTTERBUCH, 1586)

FIG.5.10: THE HORSE CHESTNUT TREE FROM *SILVA* (1776 EDITION)

In *Sylva* (1679 edition) Evelyn writes: 'I wish we did more universally propagate the Horse-chesnut, which being easily increase'd from layers, grows into a goodly Standard, and bears a most glorious flower, even in our cold Countrey.'

hard', sweet chestnut leaves on the other hand make 'very wholesome mattresses to lie on ... but for the crackling noise they make when one turns up on them, which the French call "Licts de Parliament".'

FIG.5.11: OAK TREE FROM *DELICIAE ET ARCANA FLORUM* BY ANDREAS DE LA CROYX (MAGDEBURG, 1710)

Historically throughout Europe the sudden shortages of timber have caused problems. During the Bronze Age, on the island of Cyprus, the copper industry collapsed because of the non-availability of furnace wood; likewise in the Middle Ages the bakers of Montpelier in France held a *manifestation* due to lack of fuel for their ovens. In England in 1544 Henry VIII made a proclamation regarding the long-term supply of oak [Fig.5.11]:

> Twelve Standils or Storers [young trees] likely to become Timber Trees, shall be left on every acre of Wood or Underwood that shall be felled at or under 24 years Growth; the turning of Woodland into Tillage is prohibited; and whenever any Wood is cut, it must be immediately enclosed, and the young Spring of Wood protected for Seven years; and Penalties are appointed for every Transgression against this Law.

The problem was the constant and increasing requirement for timber in all its forms; the slow but steady rise in population meant the building of more houses and making of furniture; increasing trade of itself needed wood for manufacturing items and wooden carts [Fig.5.12] and trunks had to be available for transporting them. The iron industry, glass-making, dyeing of cloth, salt panning, tin mining, all devoured wood in the form of fuel for their

FIG.5.12: A SURREY WAGON, *CIRCA* 1800

furnaces. Above all it was ship-building [Fig.5.13] which was the most vulnerable to the supply or otherwise of timber. The sixteenth century saw the beginning of overseas exploration followed by the trading and supply of goods to settlements, with the returning ships laden with raw materials which were needed to power-house manufacturing in England. It was however the navy, whose requirements were paramount. The enlargement of the fleet during the seventeenth century was huge and quite obviously caused major problems in sourcing the correct timber. In 1603 the naval fleet had a displacement tonnage of only 17,000 tons, sixty years later it had risen by over 200% to 57,000 tons. No wonder that Evelyn's book was received so warmly and read so avidly.

Sylva consists in the whole of thirty-five chapters, of which twenty-one are concerned with specific trees or groups of trees. Evelyn mentions about eighty species in total, of which under half are native, and none of which are endemic to Britain. The majority of trees he draws to our attention are introductions mostly from southern Europe and around the Mediterranean with, however, a significant proportion being discovered and shipped over from America. This balance reflects not only the paucity of native British trees, compared with some other countries, but

FIG.5.13: *THE 'ROYAL GEORGE' AT DEPTFORD SHOWING THE LAUNCH OF 'THE CAMBRIDGE'.*
JOHN CLEVELEY, THE ELDER, OIL ON CANVAS (1757)

FIG.5.14: TIMBER WAS IN DEMAND BY MANY CRAFTSMEN IN THE EIGHTEENTH CENTURY INCLUDING THE COOPER AND THE WHEELWRIGHT (BELOW).
BOTH ILLUSTRATIONS ARE FROM DENIS DIDEROT'S *L'ENCYCLOPÉDIE, OU DICTIONNAIRE RAISONNÉ DES SCIENCES, DES ARTS ET DES MÉTIERS*

the open-mindedness of Evelyn who welcomed any tree that he believed would relieve the timber crises. To grow trees and produce timber he explains, sometimes in great detail, how to sow the seeds, manage them, where, when and how to plant seedlings, when to fell, and how to fell timber. More esoterically Evelyn gives recipes for the making of varnish (and lacquer which was just arriving from Japan), the best method for the priming and painting of wood. He goes on to explain which woods were best for carving, making tools, and furniture, how baulks of specific timber were the most suitable for underwater work like the construction of locks, making of bridges or supporting river banks [Figs.5.14 & 15].

FIG.5.15: THE WHEELWRIGHT (SEE ABOVE)

Food and medicinal uses are occasionally explained, but so are the other extracted products from trees (particularly the *pinus* species), such as resin, tar, and turpentine [Fig.5.16].

The remaining fourteen chapters both support and conclude Evelyn's detailed and specific writing about his selection of trees. The five penultimate chapters mostly contain, and re-iterate, the practicalities of arboreal maintenance, but the last three chapters contain Evelyn's explanation and encouragement of why the planting and growth of trees meant so much to the ancient world, and why it should be continued. He philosophises about the impact the sacred grove had on civilisation, and why it is a duty to improve the natural world; as Evelyn reiterates time and time again, it is the planting of trees which is the central core of maintaining the balance between spirituality and practicality. Although he does not use the word, Evelyn is advancing the modern ideas of conservation and protection to the natural world.

It is this combination of practicality combined with overall vision which makes John Evelyn unique in his writing about trees. Unlike today, where forest planting and forestry work is allied to farming and industry, no such distinction took place in the seventeenth century. Evelyn's only criteria were the usefulness of the tree, alive or as timber, and whether, while growing, it could enhance its surroundings. Reading his book must have been an inspiration to many who, perhaps for the first time, saw the possibility of combining beauty with pleasure and utility with work. It was not a completely new idea, the precedent for such an idea had tentatively been set over a hundred and fifty years earlier by Henry VIII's erstwhile Treasurer and Chancellor, Sir Thomas More (1478-1535), who in his book *Utopia* (1516) had

FIG.5.16: COLLECTING SAP. *SYLVA* (1679 EDITION)

written of creating garden cities and enhancing the countryside to make it a more pleasurable space.

Although Evelyn died in 1706 his reputation and that of *Sylva* continued to make an impact. His book remained in print, in various forms, for over a hundred and fifty years, the last edition being published in 1825. His diary had been published a few years prior to that in 1818, which was about two or three years before that of Samuel Pepys. Various eighteenth- and nineteenth-century figures were well aware of John Evelyn's writings and the influence he still commanded: Lord Nelson (1758-1805) enquired why his recommendations regarding the planting of trees specifically for use by the Royal Navy had not been carried out. Robert Southey (1774-1843), the Poet Laureate, called him a 'perfect model of an English gentleman'. Two hundred years after Evelyn's death, Benjamin Disraeli (1804-81) greatly admired his erudition and his integrity.

Trying to get to grips with the supply and demand for timber always proved difficult and there were about fifteen Royal Reports or Commissions on the subject during the late eighteenth and early nineteenth centuries. Plenty of detailed information was available, and is as relevant today as it was then:

> Commerce and industry seek for, and are supported by, speedy returns of gain, however small; and the more generally the commercial spirit shall prevail in this country, the less probability is there that planting of woods for the advantages of posterity will be preferred to the immediate profits of agriculture.

A second quote seems even more appropriate to the twenty-first century:

> It is true that there is no kind of property that requires the protection of laws more than timber which may easily, and in very little time, be hurt or destroyed but requires a century to come to perfection.

The concluding of *Sylva* gives Evelyn the opportunity to express his own beliefs, when he turns from writing about the worshipping of the gods in their sacred groves to propounding the Christian faith and his own high Anglicism. He believes that Paradise itself was the ultimate sacred grove 'planted by God himself and given to Man'. This phrase, slipped into his writing without any preamble, represents Evelyn's deepest held belief and is at the core of the whole of his writings and indeed his life. It explains his idealism and almost religious fervour in his urgency for the planting and nurturing not only of trees but the whole of nature itself. It is a most fundamental but largely unexpressed ideal of the whole of humanity regardless of faith or race.

John Evelyn ends the book where he began:

> But I have done, and it is now time for us to get out of the Wood, and to recommend this, and all that we have proposed to His Most Sacred Majesty, the Honourable Parliament, and to the Principal Officers, and Commissioners of the Royal Navy.

FIG.5.17: THE CEDAR TREE FROM *SILVA* (1801 EDITION)

MAGGIE CAMPBELL-CULVER is a garden and plant historian and a Fellow of the Linnean Society. She was the Garden Conservationist at Fishbourne Roman Palace and in charge of garden restoration and the running of Mount Edgcumbe for fourteen years. In 2001 Maggie published *The Origin of Plants*; she is one of the editors for the new edition of *The Oxford Companion to the Garden* (2006), and has contributed to the Insight Guide *Great Gardens of Britain and Ireland* (2004) and to English Heritage *Handbook on Management of Historic Gardens and Parks* (2005). Maggie's paper for this conference forms part of her latest book, *A Passion for Trees, the Legacy of John Evelyn* (2006).

FIG.6.1: JOHN EVELYN IN HIS LATE 60S, PAINTED BY SIR GODFREY KNELLER IN 1689, 'HOLDING MY *SYLVA* IN MY RIGHT HAND'.
BY THEN, THREE EDITIONS HAD BEEN PUBLISHED IN 1664, 1670 AND 1679.

6

GREENHOUSES AND GREAT STORM: JOHN EVELYN ON THE WORKINGS OF GOD AND MAN IN THE GARDEN

MARK LAIRD

This essay takes up the story of John Evelyn's final years from 1685 to 1706 [Fig.6.1].[1] We find him contending first with personal loss, then finally with the loss of the nation's woods through the great storm of 26/27 November 1703 when he was preparing his fourth edition of *Sylva* for the press. The storm seemed to Evelyn, among other calamities, a 'judgement' by God. Chastising national and personal sins, God authorized it like a 'plague not to be prevented'.[2] By contrast, the greenhouse symbolized man's ability to control environment.

Reaching the '85 yeare of Age',[3] and having lost sons and daughters, Evelyn found that he had outlived even the 'goodly *Oaks* subverted by that late dreadful *Hurricane*'.[4] In old age, prayers fed off remembrance. He took the reader back to the frosts of 1683/4, which 'left such *Marks* of their *deadly* Effects'. He explained how the 'Valiant and Sturdy' oak appeared acutely prejudiced by the cold and by the drought of the ensuing year and he rounded off his account with the hurricane of November 1703, 'which begins the *Epocha* of the *Calamities*' (leading up to a '*Tempest*' around August 1705).[5] That '*dreadful* Hurricane' he linked to other storm damage including:

> that surprizing *Blast*, accompany'd doubtless with a *fiery Spirit*, which smote the most flourishing *Foresters* and *Fruit-Trees*, burning their *Buds* and *Leaves* to Dust and Powder, not sparing the very *Fruit*. This being done in a Moment, must be look'd upon as a *Plague* not to be prevented.[6]

For such 'plagues', Evelyn had in his hortulan arsenal a practical tip as well as a prayer:

> In the mean time, that the Malignity proceed no farther, it may be advisable to Cut, and Top the *Summities* of such tender *Mural* Trees, rare *Shrubs*, &c. as

FIG.6.2: S. HIERONYMUS GRIMM, THE GREENDALE OAK AT WELBECK IN 1775
FROM *SILVA* (1801 EDITION)

In *Sylva* (1679 edition) Evelyn describes the Greendale Oak: 'This Oak which stands in this [Welbeck] Lane call'd Grindal Oak has at these several distances from the ground these Circumferences: At 1 foot, 33 foot, 1 inch; at 2 foot, 28 foot, 5 inch; at 6 foot, 25 foot, 7 inch. The breadth is from bough end to bough end diametrically 88 foot; the height from the ground to the top-most bough 81 foot ... There are three Arms broken off and gone, and eight very large ones yet remaining, which are very fresh and good Timber.'

have most suffer'd, and are within reach, rubbing off the Scorchings in order to new *Spring.*[7]

Like a blasted oak regenerated [Fig.6.2], Evelyn led the final twenty years of his life with fresh starts and new springs. He might have written in his mid-60s that ''tis late for me to begin new paradises'.[8] But after the bitter winter of 1683/4, he found hope in renewal: he revised the layout plan for Sayes Court on 3 February 1685. By 1686 he was writing the 'Directions' for Jonathan Mosse [Fig.6.3], who began a six-year term as apprentice gardener on 24 June. Above all, having viewed the subterranean stoves of John Watts's new 'Conservatory' at Chelsea in August 1685, his imagination was fired anew. He published a new method of heating a greenhouse in the 1691 edition of his *Kalendarium Hortense*. Others had experimented with the solar heat of glass. Yet Evelyn saw combustion as an effective way of creating a terrestrial paradise of perennial greens and perpetual fruiting. As the great storm would confirm, recycled heat and air were part of the cycle of life. Yet so equally they were matters for human agency. The greenhouse harnessed light from the sun above, whilst light stored in matter (oak, coal, peat) could be burnt in vaults below. The doings of man went hand in hand with the workings of God.

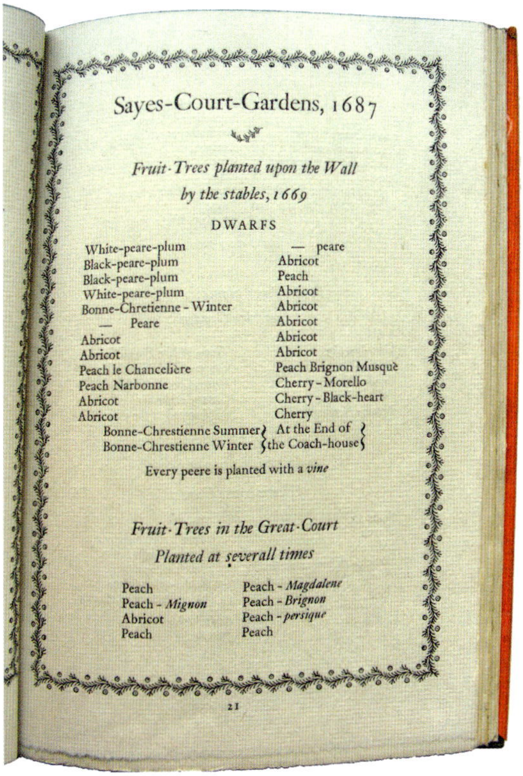
Sayes-Court-Gardens, 1687

Fruit-Trees planted upon the Wall
by the stables, 1669

DWARFS

White-peare-plum	— peare
Black-peare-plum	Abricot
Black-peare-plum	Peach
White-peare-plum	Abricot
Bonne-Chretienne - Winter	Abricot
— Peare	Abricot
Abricot	Abricot
Abricot	Abricot
Peach le Chancelière	Peach Brignon Musquè
Peach Narbonne	Cherry - Morello
Abricot	Cherry - Black-heart
Abricot	Cherry
Bonne-Chrestienne Summer } Bonne-Chrestienne Winter }	At the End of the Coach-house

Every peere is planted with a *vine*

Fruit-Trees in the Great-Court
Planted at severall times

Peach	Peach - *Magdalene*
Peach - *Mignon*	Peach - *Brignon*
Abricot	Peach - *persique*
Peach	Peach

21

FIG.6.3: PAGE FROM *DIRECTIONS FOR THE GARDINER AT SAYS-COURT*

The story of Evelyn's last twenty years should begin in 1685: a new reign. After the bitter season of 1683/4, Evelyn planted new fruit trees.[9] On the first day of that new year, 1685, he prayed for blessings and benefits, but felt a chill wind blowing: 'I implord the continuance of

Gods mercy & providence for the yeare now enter'd ... It proved so sharp weather and so long & cruel frost that the Thames was frozen crosse' [Fig.6.4].[10] By 7 March, just a month after drawing the new plan, he noted: 'Newes coming to me that my Daughter Mary was falln ill of

FIG.6.4: *A FROST FAIR ON THE THAMES AT TEMPLE STAIRS* BY ABRAHAM HONDIUS (1684)

the Small Pox, I hastned home full of apprehensions'.[11] A week later she had 'rendred [up] her soule to the Lord Jesus'.[12] She was nineteen. Evelyn was distraught beyond measure. On 24 May 1685, he turned to the bleak garden, writing:

> We had hithertoo [not] any raine for many monethes, insomuch as the Caterpillar had already devoured all the Winter fruite through the whole land, & even killed severall greate & old trees; such two Winters, & Summers I had never known:[13]

Although plentiful rain came at the end of June, the damage to the sturdy oaks was such that 'some *Ages* will hardly repair'.[14]

By 27 July that summer, another daughter, Betty, brought further distress into the desolation of her parents' home: 'This night when we were all asleepe went my Daughter Eliz: away, to meete a young fellow, nephew to Sir Jo: Tippet (Surveyor of the Navy: & one of the Commissioners) whom she married the next day Tuesday'.[15]

Just weeks later, Evelyn would enter in his diary that his 'undutifull daughter', though nursed by his 'disconsolate Wife', had died of small-pox; 'thus in lesse than 6 moneths were we depriv'd of two Children for our unworthinesse, & causes best known to God'.[16]

In August 1685, seeing the 'Conservatory' at Chelsea must have restored some faith in human agency in this 'vale of misery'.[17] The 'Tree bearing the Jesuits bark' could bring down the quartan fever of malaria. The 'subterranean heate' could overcome the coldest frosts. Evelyn, visiting other notable gardens that autumn and into the following seasons, slowly found new grace. In late October 1685, Lady Clarendon showed him around Swallowfield, where he commented on 'two very noble Orangeries well furnish'd'.[18] Delicious and rare kinds of fruit pleased him. He noted how Lady Clarendon was 'extraordinarily skilld in the flowry part', while Lord Clarendon excelled in 'the planting'. The sweet calamus of the Bible seemed to bless the household:

> The Waters are all flag'd about with Calamus arromaticus [sweet flag or *Acorus calamus*]; of which my Lady has hung a Closset, that retaines the smell very perfectly: Also a certaine sweete willow & other exotics: ...[19]

In March 1688 he was at Sir William Temple's garden: 'The most remarkeable thing, is his Orangerie & Gardens; where the wall Fruite-trees are most exquisitely nailed & applied'. He went on to Sir Henry Capel's Kew, 'whose Orangerie & Myrtetum, are most beautifull, & perfectly well kept: He was contriving very high palisados of reedes, to shade his Oranges in during the Summer, & painting those reedes in oyle'.[20]

From 1685 to 1688, as Evelyn's circumstances changed for the better, he realized that there was still garden work to accomplish. The arrival of William and Mary seemed to bring some providential blessing. On 21 April 1689 he noted in his diary: 'This was one of the most seasonable Springs, free from the usual sharp Eastern winds: that I have observ'd since the yeare 1660; at the Restauration of K. C. II: which was much such another'.[21] In July 1689 Sir Godfrey Kneller, at the request of Pepys, finished the portrait of Evelyn, which his daughter Susanna found 'very like & finely painted'.[22] The bony fingers grasp his *Sylva* emblematically [see Fig.6.1 again]. His concern for reformation in England, born of previous bitter disappointments, still depended on a prayer for human

FIG.6.5: A TRADITIONAL FREE-STANDING STOVE HEATS THIS GREENHOUSE WITH ITS TENDER TREES AND SHRUBS ARRANGED IN TUBS AND POTS.
FROM JAN VAN DER GROEN, *DEN NEDERLANDTSEN HOVENIER* (AMSTERDAM, 1683)

action: '*Good and wise government*', '*Prudent Counsels*' and '*Due Execution of Justice*' – as well as for '*A Fruitful Season*' with '*Temperate Air*'.[23]

Tempering the air for the benefit of fruit had always been the ideal of the greenhouse. Citrus fruit [Fig.6.6], cultivated in orangeries in the Low Countries from about 1600, only required winter protection at around 8°C [Fig.6.5]. Thus, when Evelyn first encountered the novel idea of growing tropical shrubs and trees in a 'stove' at Chelsea, he noted the key innovation was 'subterranean heate'. John Watts seems to have learned this from Paul Hermann at Leiden on his visit in 1683.[24] A second innovation came from the Netherlands. Glass frames placed on hotbeds had already demonstrated the power of the sun's rays on the shortest days of the year. Hence it was logical to enlarge the frames into 'glass houses', using the additional heat of underground furnaces. While the credit for this invention remains unclear,[25] the innovation shows up in the Hortus Medicus in Amsterdam by 1682 or 1683 [Fig.6.7]. An engraving of the 1682 layout illustrates two narrow structures. The larger one, facing south, was 80 feet long and only 4 to 5 feet wide. A smaller one was 36 feet long and 5 feet wide.[26]

The same idea lies behind three 'glass cases' at Hampton Court that were built for Queen Mary by the Dutch carpenter Hendrik Floris. A drawing in the 'Fort Album' of *circa* 1718 records one of the glass cases erected in 1689 [Fig.6.8]. Christopher Hatton described them as 'much better contrived and built than any other in England'.[27] By August 1691 Mary Evelyn was writing to her husband that the stove was: 'filled with the most rare plants of the Indies[,] things wonderfull in

FIG.6.6: SEVILLE ORANGE.
DETAIL FROM A PAINTING BY
ALEXANDER MARSHAL (*CIRCA* 1650-82)

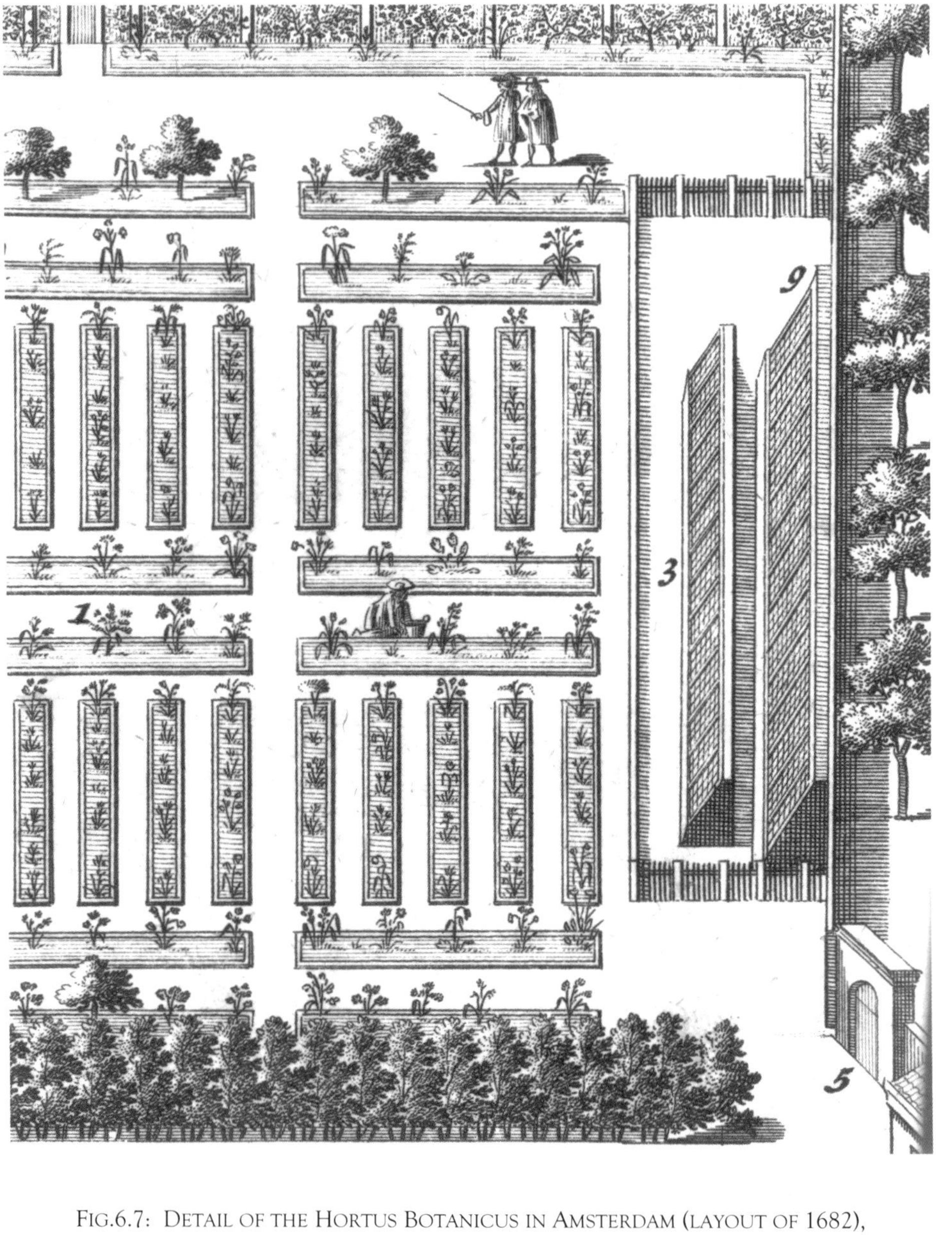

FIG.6.7: DETAIL OF THE HORTUS BOTANICUS IN AMSTERDAM (LAYOUT OF 1682), ENGRAVED BY B. STOOPENDAAL (1685).
FROM C. COMMELIN, *BESCHRYVINGE VAN AMSTERDAM* (AMSTERDAM, 1693)

their kind some leaves three quarters of a yard long and a quarter broad others of different forms not to be described'. Evelyn had become aware of glass stoves as he began to think about heating greenhouses.

The 1691 edition of Evelyn's *Kalendarium Hortense* reflects the thinking he had done since viewing the 'subterranean heate' at Chelsea in 1685:

> 'Tis now after many severe *Winters* Observation, both whilst they made use of ordinary *Iron Stoves* and other Inventions to moderate the sharp *Air* in the *Green-house* (as they call it) and even since the *Subterranean Caliducts* have been introduc'd, I often took notice that ... *Plants* ... rarely pass'd their Confinements, without Sickness, a certain *Languor* or Taint discoverable by their Complexions.[28]

The plants could not breathe, as he presumed, 'the pure and genuine *Air*, impregnated with its *Nitrous Pabulum*, which is not only the Nourishment, and Life of *Animals*, but of all *Plants* and *Vegetables* whatsoever'. The 'pent-in' air was effete and vitiated.[29]

Hence Evelyn came up with the idea of having the stove outside the house. He visualized the heat being transferred inside by hot pipes. In turn, cold air from the floor area was drawn through a ground pipe to fan the flames of the furnace. Natural convection would therefore ensure a continual supply of fresh and warm air. However effective in producing circulation, this method was probably less efficient in thermal terms.[30] The diagrams in

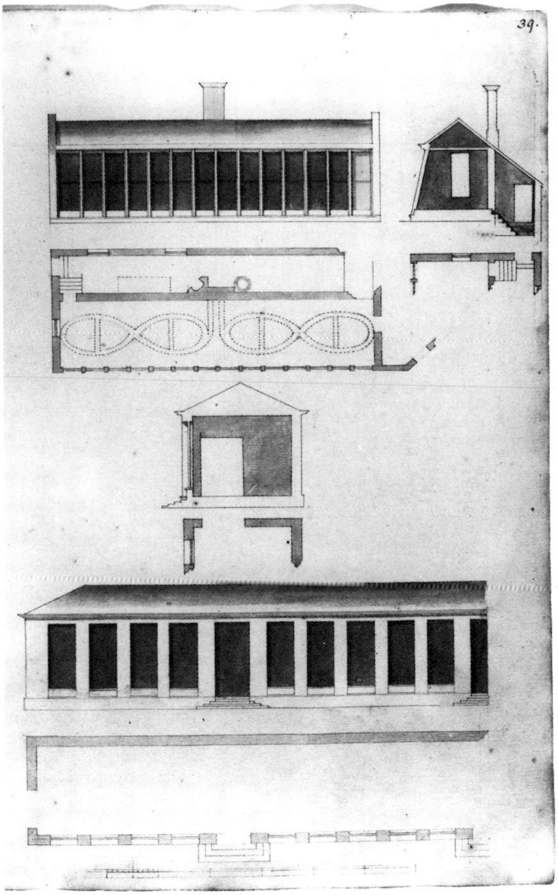

FIG.6.8: HAMPTON COURT, ONE OF THE GLASS CASES OF 1689 (ABOVE) AND THE ORANGERY OF 1701 (BELOW). FROM THE 'FORT ALBUM' (*CIRCA* 1718)

Kalendarium Hortense are detailed to good effect [Fig.6.9]. Those labelled 'Figures II, III and IV' are details of the general 'Figure I'. The latter shows the whole greenhouse with its exterior '*Stove* or *Furnace*'. As with a conventional hearth, the ash is collected below (1) the fire itself (2). The chimney (4) serves to emit the 'exhausted *Air*' of the greenhouse as well as the smoke from the stove. That stale air comes through under-floor piping (5) from an aperture (6), and serves to fan the flames of the fire. Although a thermometer is placed on the end wall (7), it allowed only a subjective reading of heat. Effective thermometers were first introduced after 1714.

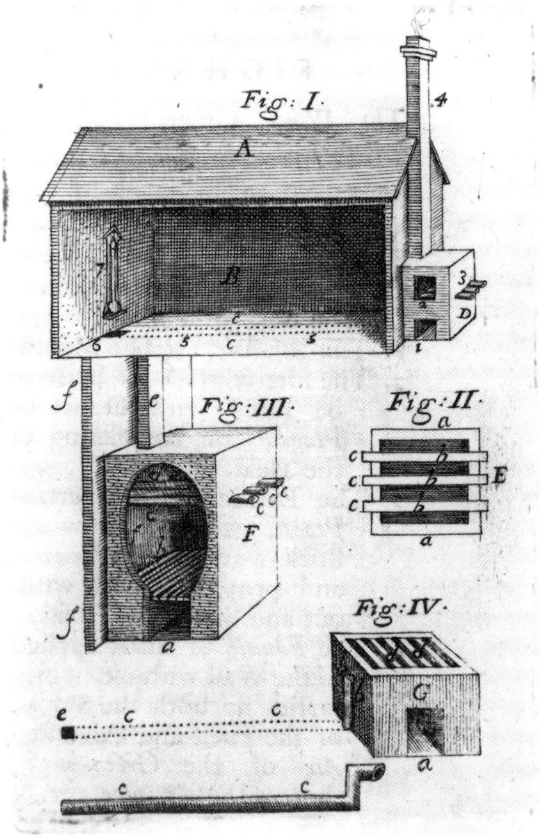

FIG.6.9: JOHN EVELYN'S DESIGN FOR A GREENHOUSE WITH EXTERNAL FURNACE.
FROM *KALENDARIUM HORTENSE* (LONDON, 1691)

Refinements followed as the idea of the external stove was tested in practice. From Evelyn's exchange of letters with Sir Dudley Cullum,[31] it is clear that Archbishop Tillotson's experiment of 1691 at Lambeth Palace needed correction and improvements. Pipes, whether of iron or clay, were best kept away from the red-hot coals, argued Cullum. Evelyn agreed in his reply of 14 February 1692, adding that, to distribute the heat effectively from the hearth, clay pipes might be of wider gauge or placed along the wall with 'overtures' to let the 'vapour' out. Dudley Cullum laid his 19"-pipes – 'made of Crucible earth' – sixteenth inches above the fire. He introduced iron plates in his paving. By opening and shutting them, he could regulate the supply of air to the hearth. And he installed wall pipes with apertures. Whether he lined the walls with mattresses or reeds – as was common – is not divulged in the letters. Evelyn had always favoured cork.[32]

The effectiveness of the system would be tested out in the bitter winter of 1694/5. It followed a glorious summer that left the garners full of grain.[33] Cullum could report in his letter of 30 May 1694 that 'notwithstanding the closeness of the place (rendring it so darke when all was shut up that I could not see my hand), yet by the Admission of that Aire you Advis'd ... one would have thought himself abroad in the open Aire in Aprill, when in January all things then without doores (nay in the house) were freezing very hard'.[34] While many plants itemized – citrus [Fig.6.10], myrtle and oleander – were those of the heated greenhouse, a succulent 'Green-Sedum' and tropical 'Indian Jessamines' also survived in the same space. This undifferentiated approach appears to lag behind the specialized glass hothouses for tropical plants alone, such as the 'self-contained' one at Hampton Court. But the advancement was clear enough all the same.

Thus what John Beale had called 'advancing the Light of nature into the Light of grace'[35] gradually began to fulfil Evelyn's desideratum: the perfection of horticulture. Evelyn understood the need for windows to be large and ample 'for *Light* it self, next *Air*, is of wonderful importance'.[36] Glass was becoming cheaper, thinner and finer. The size of a spun disc could reach several feet, though 18 inches was the largest standard pane. By 1695 King William imposed a very high duty on glass, which was only repealed in 1698. The three years were notable for the lack of new greenhouse construction.[37]

Employing diverse technologies and hands-on care, gardeners were succeeding for the first time with the larger woody species of the tropics. In the one case, a solar greenhouse effect was being exploited (and oak trees were sacrificed for banana trees [Fig.6.11]). In the other, fossil-fuel combustion produced perpetual spring.

FIG.6.10: ORANGE TREE FLOWER BY ROBERT NICOLAS (1614-85) FROM *LA GUIRLANDE DE JULIE*, CIRCA 1642

Hence, in short, an Elysium on earth seemed almost at hand. However, in late November 1703 destruction came from on high. John Evelyn's hurricane entry in his diary for 26/27 November evokes the sense of calamity in the whole of southern England, and by the example of Wotton [Fig.6.12]:

> The dismall Effects of the Hurecan & Tempest of Wind, raine & lightning thro all the nation, especial (y) London, many houses demolished, many people killed: [27] & as to my owne losse, the subversion of Woods & Timber both left for Ornament, and Valuable materiall thro my whole Estate, & about my house, the Woods crowning the Garden Mount, & growing along the Park meadow; the damage to my owne dwelling, & Tennants farmes and Outhouses, is most Tragicall: not to be paralleled with any thing hapning in our Age (or) in any history almost, I am not able to describe, but submitt to the Almight(y) pleasure of God, with accknowledgement of his Justice for our National sinns, & my owne, who yet have not suffered as I deserved to: Every moment like Jobs Messengers, bring(s) the sad Tidings of this universal Judgement.[38]

FIG.6.11: EVELYN'S DRAWING OF A BANANA TREE FROM 'ELYSIUM BRITANNICUM'

He would cite the public reckoning in his 1706 edition of *Sylva*: 'no less than 3000 brave *Oaks*, in one part only of the *Forest* of *Dean* blown down'. He totted up cumulative figures: 'in about 450 *Parks* and *Groves*, from 200 large Trees to a 1000 of excellent Timber, without counting Fruit and Orchard Trees *sans* number, and proportionably the same through all the considerable *Woods* of the Nation; with those stately *Groves*, *Avenues* and *Vista's*'.[39] Doubtless the true figures ran close to the millions that the great storm of October 1987 laid low [Fig.6.13].[40]

When he came to write on Wotton, the account goes beyond the ledgers: 'above 2000' trees blown down, some raising 'Mounds of Earth near 20 foot high, with great Stones intangled among the Roots and Rubbish'.[41] As he recalled elsewhere in *Sylva*:

> In a word, to give an *Instance* of what store of *Woods* and *Timber* of prodigious size, there were growing in our little County of *Surrey*, ... (with sufficient Grief and Reluctancy I speak it) my own *Grandfather* had standing at *Wotton*, and about that Estate, *Timber* that now were worth 100000 *l.* Since of what was left my *Father*, (who was a great *preserver* of *Wood*) there has been 30000 *l.* worth of *Timber* fallen by the *Ax*, and the *fury* of the late *Hurricane* and *Storm*: Now no more *Wotton*, stripped and naked, and ashamed almost to own its Name.[42]

The ecological decline of woodlands in the Tillingbourne valley in Surrey had begun before the Evelyns settled at Wotton. Then in successive generations, reckless cutting for fuel and manufacturing – combustion for gunpowder or brass – shifted common woodland to encoppiced holdings and heathland.[43] The hurricane merely compounded this degradation. For the 'wood-born' and conservation-minded Evelyn, his lost 'wood-town' was woe enough: 'sufficient to mortifie and change my too great Affection and Application to this *Work*'.[44] Yet when he finally got to Sayes Court on 10 February 1704, the remembrance of

FIG.6.12: JOHN AUBREY, WATERCOLOUR OF JOHN EVELYN'S FAMILY HOME AT WOTTON. BODLEIAN LIBRARY, OXFORD.

things past must have inflamed his new grief. As he wrote in *Sylva*:

> methinks I still hear, and am sure feel the dismal Groans ... of our *Forests*, so many thousands of goodly *Oaks* subverted by that late dreadful *Hurricane*; prostrating the Trees, and crushing all that grew under them, laying in ghastly Postures, like whole *Regiments* fallen in *Battle*, by the Sword of the *Conqueror*: Such was the Prospect of many Miles in several Places, resembling that of *Mount Taurus*, so naturally describ'd by the *Poet*, speaking of the Fall of the *Minotaurs* slain by *Theseus*.[45]

Following the 'publique fast after the dreadfull storm' of 19 January 1704, storm sermons were preached (and printed) on that day throughout the reign of Queen Anne. Even into the 1730s, in accordance with an endowment, the Lord's mercy was invoked in storm sermons at the Congregational Church in Little Wild Street. William Talbot, Bishop of

FIG.6.13: AFTER THE 1987 STORM.
WOODLAND ON THE FORMER SHABDEN PARK ESTATE, CHIPSTEAD.
PHOTOGRAPH BY IAN MOORHOUSE FROM *SURREY IN THE HURRICANE*

Oxford, speaking in Westminster Abbey on 19 January 1704, made his sermon a matter of judgement: 'the Lord was in the Wind ... it was Appointed by Him ... whose Word the Storms and Tempest fulfil, to awaken and chastise ... a sinful people'.[46]

For all this, Evelyn had enjoyed many quiet moments with his wife at Wotton – more than for fifty years past, he admitted.[47] He continued writing and planting. Redemption was possible, as earth foretold and as heaven promised. Even after the subversion of oaks in thousands on 26/27 November 1703, 'a most famous and monstrous *Oak* growing at *Epping* in *Essex*, (blown down) raised itself, and withstood that *Hurricane*'.[48] Despite the 'dismal Groans of our forests', and despite the self-castigating acceptance of a judgement of national and personal sins, he could end *Sylva* on an ascendant note:

> [I] shall if God protract my years and continue health be continually planting 'till it please him to transplant me into the glorious regions above, the celestial Paradise, planted with perennial groves and trees bearing immortal fruit.[49]

And, if the prayer that he included in the fourth edition of *Sylva* amounts to a general supplication, contentment was one desideratum:

> *That there be no Decay; no leading into Captivity:;*
> *no Complaining in our Streets:*
> *But that every Man may sit under his own Vine,*
> *and under his own Fig-tree, in thankfulness to Thee;*
> *Sobriety, and Charity to his Neighbour;*
> *and in whatsoever other Estate, thou wilt have him,*
> *therewith to be contented:*[50]

Between the pages on scripture, final entries in the diary prove testament to his enduring pleasure in gardens and natural knowledge. Despite the bodily failings, and amid drowsiness at sermons, he still found renewed gusto in the sublunary world. On 18 May 1705 he noted: 'I went to see Sir J. Jardine at Turnham Greene, the Gardens being very fine & exceedingly well planted with fruit'.[51] On 2 September (after two days of physical 'Torment'), he saw the new gardens at Kensington. He met Henry Wise, 'who was there on purpose to receive me'.[52] The perfection of horticulture offered one salvation for the here and now, though gardens declined too. Visiting Beddington on 20 September

1700, he commented on how those first orange trees in England were 'planted in the open ground, & secured in Winter onely by a Tabernacle of boards, & stoves, removable in summer'. The trees that stood for 120 years 'laden with fruite' were 'now in decay ... the negligence of a few years, ruining the Elegances of many'.[53] But the science of glass and the technology of heat had finally made perpetual spring quite possible in winter.

Despite his faith in combustion, it was solar energy that preoccupied him towards the end. He discovered how glass could use the sun's heat to generate and transmute. As he wrote on 22 April 1705

> I after many moneth(s) indisposition, since I came to Towne: went to the meeting of the R. Society ... I also saw Sir Isaac Newtons ... burning Glasse which dos such wonders ... it penetrating Cast Iron of all Thicknesse, vitrifies Brick, mealts all sorts of mettals in a moment.[54]

He went back on 17 June 1705 to look at the effect on slate, which withstood the heat longer than metal. The summer was by then turning 'exceedingly dry & hot':[55]

> The Glass was composed of 7 round burning glasses of about a foote diameter, so placed in a frame, as to cause all their Sun-beams to meete in one *focus* onely: Excessive hot & (dry), such a long time, as all the Country was burnt up.[56]

The greenhouse was not the only instance of fire and glass co-joined. Back in 1685, Evelyn had viewed combustion within glass as a link between body and spirit. In a 'noble Experiment', he saw how the 'matter of *Phosphorous*, ... made out of human blood and urine', could elucidate the 'Vital flame ... in Animal bodys'. After dinner with Pepys, as he recorded in the diary: 'Dr. *Slayer* shew'd us an Experiment of a wonderfull nature'. Liquors were poured into a 'Matras'. Then, inside that glass distilling-vessel, clouds and flames materialized. When shaken together, the liquors:

> ... fixed divers sunns & starrs of real fire perfectly globular upon the walls of the Glasse to our greate astonishment, & which there stuck like so many Constellations burning most vehemently, & exceedingly resembling starrs & heavenly bodys, & that for a long space: It seem'd to exhibit a Theorie of the eduction of light out of *Chäos*, & the fixing or gathering of the universal light into luminous bodys.[57]

Evelyn had grown up proud of his family's association with gunpowder and manufacturing in the Tillingbourne valley: alder [Fig.6.15] burnt for charcoal [Fig.6.14], pyrites for sulphur, and the chemistry of saltpetre.[58] It seems fitting that, refining combustion for the

FIG.6.14: CHARCOAL BURNING FROM *SYLVA*, 1679 EDITION

greenhouse, he was pulled in later years – like a moth to a flame in a glass – by the power of incandescent matter. In a similar light, during that last hot summer, Evelyn would have a long conversation with 'Dr. *Dikinson* the famous Chymist ... about the Philosophers Elixir, which he believed attainable'.[59] Such alchemy of metals made into gold appears an idle dream. Yet through practical hortulan technologies, Evelyn sustained a genuine faith in regaining a 'lost felicitie', especially by means of the garden. Resisting storms from on high, his heated greenhouse came closest to man's re-creating Eden in England.[60]

Today, scientists look at the links of 'pent-in' fossil-fuel emissions – the greenhouse effect – to the carbon cycle and soil carbon.[61] They consider the impacts of climate change on forest vegetation, including the woods of Wotton that have endured centuries of exploitation.[62] They are also concerned with

FIG.6.15: ALDER CATKINS, FROM *SILVA*, 1801 EDITION

the hurricane link to climate change.[63] Nowadays, most of us have lost faith in the prayers of Evelyn's day to ward off 'acts of God'. We look instead to governments and legislation to mitigate environmental degradation. Yet, in the workings of God and man in the garden, Evelyn prayed for good government and effective laws as much as for providential blessings. Evelyn's call in *Sylva* to protect and cherish trees, compromised as it may appear by the family's association with resource exploitation and explosives, is particularly resonant in our own conflicted times as we celebrate the 300th anniversary of his death.

In memory of Teige O'Brien of Painshill Park Trust in Surrey.

I would like to thank Mavis Batey for her generous help in pruning this essay, which is an abbreviated account of the paper that first appeared in *Garden History*, vol. 34, no. 2 (2006).

FIG.6.16: OXFORD PHYSIC GARDEN, THE HEATING APPARATUS USED UNTIL ABOUT 1837, WHICH WAS SIMILAR TO THE 'WAGON' FILLED WITH OAK LOGS AT HAMPTON COURT

REFERENCES

1 For an account of Evelyn at Sayes Court prior to 1685, see Mark Laird, 'Sayes Court Revisited', in Frances Harris and Michael Hunter (eds.), *John Evelyn and His Milieu* (London, British Library, 2003), pp. 115-144 and, especially, pp. 138-140.

2 John Evelyn, *Sylva* (London, 1706), p. 200. The published title was given as *Silva*.

3 E.S. de Beer (ed.), John Evelyn's *Diary*, vol. 5, p. 614. The date was 31 October 1705.

4 Evelyn, *Silva* (1706), p. 341. (NB: the pagination repeats itself over several pages, so that p.341 here refers to what should be p. 349).

5 All phrases in this paragraph come from *Silva* (1706), p. 200. The text reads '*Sept.* 1703', but the great storm was 26/27 November 1703. It is not clear what the 'tempest' around August 1705 refers to, though his diary contains many entries on the heat of that summer.

6 Evelyn, *Silva* (1706), p. 200.

7 Ibid., p. 200.
8 Quoted from Frances Harris, *Transformations of Love: the Friendship of John Evelyn and Margaret Godolphin* (Oxford, Oxford University Press, 2002), pp. 299-300. She provides a good summary of Evelyn's circumstances in 1685, and discusses the move from Sayes Court to Wotton in 1694.
9 For a discussion of the fruit-planting of 1685, see Prudence Leith-Ross, 'Fruit Planted around a New Bowling Green at John Evelyn's Garden at Sayes Court, Deptford, Kent, in 1684/5, *Garden History*, vol. 31, no. 1 (Spring 2003), pp. 29-33.
10 Evelyn, *Diary*, vol. 4, p. 402.
11 Ibid., p. 420.
12 Ibid., p. 420.
13 Ibid., p. 446.
14 Evelyn, *Silva* (1706), p. 200.
15 Evelyn, *Diary*, vol. 4, pp. 460-462. By the 'onely remaining Child' he meant Susanna.
16 Ibid., pp. 463-464.
17 Ibid., p. 462 and p. 464. The visit to Chelsea was on 6 August 1685.
18 Ibid., pp. 480-481.
19 Ibid., p. 482. For the 'sweet calmus' of the Bible – an ingredient in Moses's holy anointing oil – see Exodus: 30:23.
20 Evelyn, *Diary*, vol. 4, p. 576.
21 Ibid. p. 636.
22 See Carol Gibson-Wood, 'Susanna and Her Elders: John Evelyn's Artistic Daughter', in *John Evelyn and His Milieu*, p. 245.
23 Evelyn, *Silva* (1706), p. 201. See also Steven Pincus, 'John Evelyn: Revolutionary', in *John Evelyn and His Milieu*, pp. 207-211.
24 See D.O. Wijnands, 'Hortus auriaci: the gardens of Orange and their place in late 17th-century botany and horticulture', *Garden History*, vol. 8, nos. 2 & 3 (1988), p. 74.
25 See Marisca Sikkens-De Zwaan, 'Magdalena Poulle (1632-99): A Dutch Lady in a Circle of Botanical Collectors', *Garden History*, vol. 30, no. 2 (2002), pp. 212-214
26 Wijnands, 'Hortus auriaci', pp. 75-76, 276. See David Jacques and Arend Jan van der Horst, *The Gardens of William and Mary* (London, Christopher Helm, 1988), p. 179.
27 Quoted from ibid., p. 179.
28 John Evelyn, *Kalendarium Hortense* (London: 1691; and here in the 1699 edition), p. 153.
29 Ibid., pp. 154-155.
30 See May Woods and Arete Swartz Warren, *Glass Houses: A History of Greenhouses, Orangeries and Conservatories* (London, Aurum Press, 1988), p. 31.
31 These are discussed in some detail in Douglas Chambers, 'John Evelyn and the Invention of the Heated Greenhouse', *Garden History*, vol. 20, no. 2 (1992), pp. 201-205. I have quoted from his transcriptions.
32 Woods and Warren, *Glass Houses*, p. 31.
33 Evelyn, *Diary*, vol. 5, p. 186: 'Never more glorious & steady Summer weather, Corne & all other fruits in extraordinary plenty generaly'.
34 Quoted from Chambers, 'Heated Greenhouse', p. 205.

35 Quoted from ibid., p. 205. John Beale envisaging botanical collecting in America (completing the re-creation of Eden), wrote: 'our Savior hath built a fairer Temple than that of Solomon ... Hee hath sent out the wise and learned, and his ship is landed in the other world. And this is the First, or most publique attempt that hath beene of advancing the Light of nature into the Light of grace, that all usefull arts may be completed in the great prophesy, Multi pertransibant et augebitur scientia'.

36 Evelyn, *Kalendarium Hortense* (1699), p. 157.

37 For this background, see Woods and Warren, *Glass Houses*, p. 28 and p. 36.

38 Evelyn, *Diary*, vol. 5, p. 550.

39 Evelyn, *Silva* (1706) p. 341 (349).

40 See Martin Brayne, *The Greatest Storm* (Thrupp, Gloucestershire, Sutton Publishing, 2002), p. 125.

41 Evelyn, *Silva* (1706), p. 341 (349).

42 Ibid., p. 301.

43 For a good review of woodland management in Evelyn's part of Surrey, see P.F. Brandon, 'Land, technology and water management in the Tillingbourne valley, Surrey, 1560-1760', *Southern History*, vol. 6 (1984), pp. 75-103 and, specifically in this context, pp. 91-99.

44 Evelyn, *Silva* (1706), p. 341 (349). See also Brandon, 'Land, technology and water management', pp. 97-98, in which he cites Evelyn's *Memoires for my grand-son* of 1704, urging his young successor to avoid wood fuel in favour of peat or sea-coal in all but the living room fires.

45 Ibid., p. 341 (349).

46 Quoted from Brayne, *Greatest Storm*, pp. 211-212.

47 See Harris, *Transformations*, pp. 299-300.

48 Evelyn, *Silva* (1706), p.199.

49 Quoted by Harris in *Transformations*, p. 300, citing *Sylva* (1706), ed. John Nisbet (1908), vol. 2, p. 260.

50 Evelyn, *Silva* (1706), p. 201.

51 Evelyn, *Diary*, vol. 5, p. 594.

52 Ibid., pp. 606-607.

53 Ibid., p. 427.

54 Ibid., p. 592.

55 Ibid., p. 599.

56 Ibid., p. 600.

57 Evelyn, *Diary*, vol. 4, p. 491.

58 Brandon, 'Land, technology and water management', 79-80. See also Steven Pincus, 'John Evelyn: Revolutionary', p. 198, and Isabel Sullivan, 'Their "Own Sweet Country": The Evelyns in Surrey', in *John Evelyn and His Milieu*, pp. 281-84.

59 Evelyn, *Diary*, vol. 5, p. 599.

60 See again Laird, 'Sayes Court Revisited', pp. 138-140.

61 See, for example, Dennis Baldocchi, 'The carbon cycle under stress', *Nature*, vol. 437, 22 September 2005, pp. 483-494; Ph. Ciais *et al.*, 'Europe-wide reduction in primary productivity caused by the heat and drought in 2003', *Nature*, vol. 437, 22

September 2005, pp. 529-533; and Pat H. Bellamy *et al.*, 'Carbon losses from all soils across England and Wales 1978-2003', *Nature*, vol. 437, 8 September 2005, pp. 245-247.

62 Brandon, 'Land, technology and water management', pp. 91-99.

63 See Quirin Schiermeier, 'Hurricane link to climate change is hazy', *Nature*, vol. 437, 22 September 2005, p. 461. This short piece refers to articles in *Nature* and *Science* that argued a persuasive link.

MARK LAIRD is Senior Lecturer in the History of Landscape Architecture at the Graduate School of Design, Harvard University. He is also Associate Director of Painshill Park Trust in Surrey. Educated at the universities of Oxford, Edinburgh, and York, he was Research Fellow at Chelsea Physic Garden in London, twice a Fellow at Dumbarton Oaks, and has taught at the University of Toronto.

Based in Toronto as a consultant in historic landscape conservation, he advises on sites in Europe and North America. Recent projects include Hestercombe in England, Fürst-Pückler-Park in Germany, the Belvedere Garten and Schloß Hof in Austria, and the Halifax Public Gardens in Canada. For his replanting work at Painshill, Laird was joint recipient of a 1998 Europa Nostra medal. His research on eighteenth-century planting is presented in *The Flowering of the Landscape Garden* (1999). Using watercolour reconstruction to complement textual analysis, Laird has also published extensively on the history of horticulture from the Baroque to Gardenesque, and on conservation philosophy and practice. He is currently working on a new book on the environment of the English pleasure ground from 1650 to 1800.

FIG.6.17: BY JOHANNIS URSINI, *ARBORETUM BIBLICUM* (NUREMBURG, 1699)

FIG.7.1: THE COVER OF THE *SHELL SHILLING GUIDE TO SURREY* (1963), ILLUSTRATED BY S.R. BADMIN. THIS WAS 'EVELYN COUNTRY'

7

'INVEST IN GARDENING, GROVES AND WALKES': HOW EVELYN'S GRANDSON CARRIED OUT HIS INSTRUCTIONS TO REVIVE THE WOTTON LANDSCAPE

PETER BRANDON

'Restore the name of Wotton, otherwise in danger of being forgotten'. This was the injunction of the diarist John Evelyn to his grandson and heir who inherited the estate at the age of 23 in 1706. Few persons can have been trained so obsessively in the role of estate owner in the family tradition than Sir John and few can have so faithfully carried out the trust imposed on him. He not only re-adorned Wotton with economically profitable woodland but also brought the estate landscaping in line with the changes in taste of the 1730s and 40s.

Sir John Evelyn was the child of one of the diarist's ailing sons whose death at an early age was an acute disappointment. His brother's male children all pre-deceased their father. The diarist inherited Wotton from his brother in 1699 and his heir, then still in his teens, was brought up to bring renewed honour to Wotton.[1]

The diarist publicly acknowledged his anxieties about his inheritance of Wotton. In revising the fourth edition of his *Sylva*, published immediately on his death in 1706, he frankly informed his readers in a carefully guarded and moving statement. 'To give an instance,' he writes, 'of what store of woods and timber of prodigious size there were growing in our little county of Surrey (with sufficient grief and reluctancy I speak it), my own grandfather had standing at Wotton and about that estate, timber that now were worth £100,000. Since of what in life my father, who was a great preserver of wood, there has been £30,000 worth of timber fallen by the axe, and the fury of the late hurricane and storm. Now no more, Wotton stripped and naked, and ashamed almost to own its name'.[2]

The condition of the Wotton estate at the time of the diarist's inheritance arose from his brother's ranging of the axe 'farther than at first designed' to pay off some of his heavy debts and to provide a marriage jointure for one of his daughters. The diarist was also obliged to fell further trees to provide generous financial provision for his brother's eldest daughter and grandchildren which arose from an unexpected family dispute.[3] An indication of how the depleted woodland had detracted from the value of the estate both economically and aesthetically is in one of Evelyn's friends remarks before the Great Storm: 'I am extremely sorry the greed of some people hath driven you to cut away part of the charming grounds that made Wotton so delicious a seat. What, are those woods behind ye house towards Leith Hill cut down? If they are, the greatest ornament of the finest country in England is gone'.[4] It is one of the great ironies in the history of afforestation in England that John Evelyn whose proud boast it was that he had cultivated a universal tree sense and encouraged all to plant, protect and cherish trees everywhere, through his *Sylva*, could save Wotton in its entirety but only at the expense of its woods. This situation at Wotton caused John Evelyn acute embarrassment and humiliation. Stephen Switzer, the garden designer, likened Evelyn to another Virgil 'appointed for the relieving of calamities of England and the re-awakening of the spirit of his countrymen for their planting and sowing of woods'. Evelyn now found himself the possessor of an estate bearing all the marks of devastation which he had publicly deplored.

It was not only the Wotton woods which had decayed; but the kitchen gardens had also fallen into considerable neglect during the diarist's brother's last years. This brought additional private humiliation to John Evelyn. Only recently he had supervised the translation by London and Wise of La Quintinye's book on fruit and vegetable cultivation [Fig.7.2], *The Compleat Gard'ner*, and written *Acetaria*, a book on salads, both, as John Bowle has written, advancing 'the amenities of many a mansion, manor house, farm and rectory'.[5] The irony was not lost on Evelyn that despite these publications his own family went short of garden produce.

In these circumstances, the grandson's education at Eton and Oxford became the diarist's main concern. Even in his early teens the boy was urged to save presents of money 'towards the new building of Wotton'. The diarist obtained through his interest with Lord Godolphin, the Lord

Treasurer, the office of Treasurer of the Stamp Duties for his grandson at a salary of £300 a year. In 1705 he married him into the family of Lord Falmouth, the new Treasurer to Queen Anne, which provided the young man with necessary patronage. In 1704 Evelyn had begun to compile instructions on estate management for his grandson. In his *Memoires for my Grand-son,*[6] intended as an interim planning strategy for the revival of Wotton, the young man is reminded that the growing of timber was the most appropriate land-use on the estate on account of thin chalk soils and poor sands and, because of financial restraints, frugality was essential in every aspect of domestic and estate management. To this end he recommends that servants should bring something that has broken to show the necessity for its repair. He suggests that everything should be put under lock and key, else servants will squander them and 'sell them before your very face'. He urges him to get his wife to make regular checks on linen and household utensils 'to the very garrets' so as to restrict embezzlement and waste. Does this throw light on general relations between servants and masters in seventeenth- and early

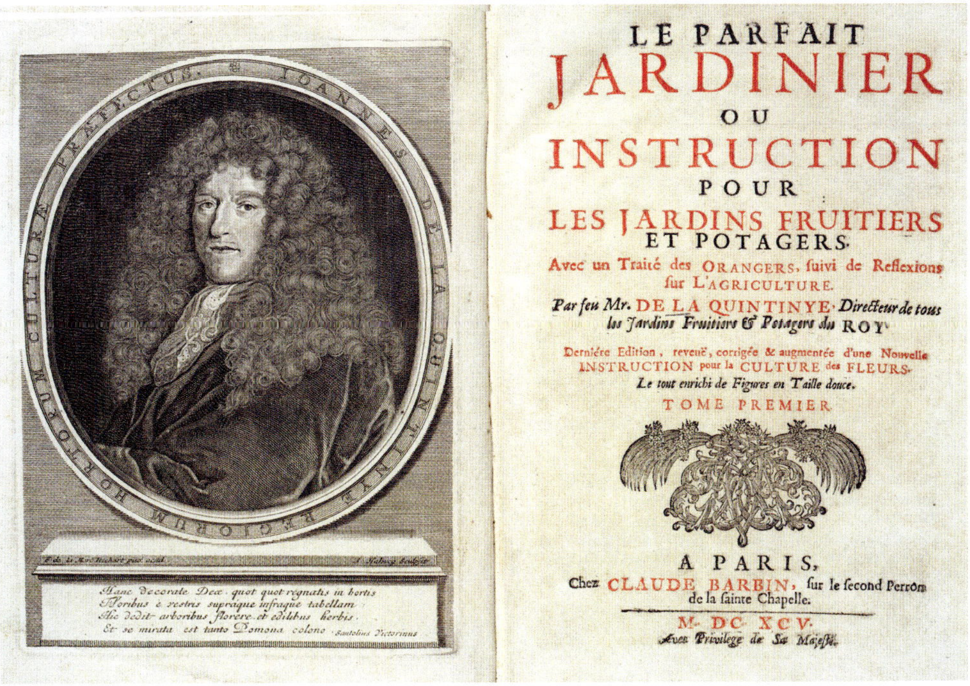

FIG.7.2: *LE PARFAIT JARDINIER OU INSTRUCTION POUR LES JARDINS FRUITIERS ET POTAGERS* WAS PUBLISHED BY JEAN DE LA QUINTINYE IN 1690. HE WAS GARDENER TO LOUIS XIV.

eighteenth-century England or did it arise from the peculiar circumstances at Wotton? As to the estate, Evelyn urges the avoidance of the unnecessary waste of wood fuel which he considered should be confined to living rooms, leaving the working rooms of kitchens, washhouse and brewhouse supplied with peat from the estate or sea-coal.

Fortunately, the grandson never betrayed the trust placed upon him. By training, inclination and temperament he came to share in most of his grandfather's intellectual enthusiasm and he inherited much of his personality, notably a natural diffidence of manner and lack of ostentation.[7] Everything he did at Wotton – his practical forestry, his building and landscaping, the keepership of the diarist's library and manuscripts, and his scientific and archaeological interests – reflects with the intensity of a mirror his deep attachment to his grandfather and a life-long concern to do his best for Wotton.

Although his life, work and character has been totally eclipsed by his distinguished grandfather, the central figure in seventeenth-century gardening and planting, his life's work brought Wotton renewed acclaim through his tree plantings and landscape design. Garden historians should not underestimate the contribution he made to the revolution in taste which in his day was pioneering forms of naturalistic landscaping.

With frugality the grandson survived the lean years when he had little money to run the estate until his sales of timber began to alter the situation some twenty years later. Although comprising 7,500 acres the estate was mainly on poor soils. The low rents of the first two decades of the eighteenth century went against him and he found it difficult at first to maintain the gardens and grounds in good repair, still less to undertake much in the way of necessary improvements. The loss of his junior government post on the death of Queen Anne obliged him to reduce expenditure still further. At this difficult time he wrote to his gardener expressing the hope that he would be 'very careful with his seeds and not want any more this great while, what I have now sent coming to a pretty deal of money'. A little later he ordered his bailiff to 'reduce the expense of the garden as much as is convenient with its being kept in pretty good order'. His financial stringency seems to have put paid to the lavish Brompton Park style of London and Wise (both had enjoyed the patronage of the diarist). Nevertheless the latter sent him plants from Hampton Court and London came twice yearly as a consultant and had sent a skilled assistant to prune the fruit trees, who was conveyed by mail coach to Epsom, the remainder of the journey being made in an estate vehicle. The bailiff was instructed to find out his diet and offer what the house affords, together with some strong beer now and again. London's imminent arrival led to the order that he was to be treated in the best manner the bailiff was able. Evelyn had to curtail expenditure later. Nevertheless a weekly hamper, bursting with produce and also containing ready cash secreted amongst it to elude highwaymen, together with a letter giving details of estate work carried out in the previous week, made its regular arrival at the Evelyns' London house. In these early years Sir John re-laid gravel walks, replaced elms in some of the walks, restored broken-down walls, re-stocked the garden with fruit trees, roses [Fig.7.3] and lilacs and attended to the Mount which had suffered in the great gale.[8]

Meanwhile, Sir John's primary concern was planting up the depleted estate, and in this he proved to be a practical woodman of the first order, transforming the estate rather like Lord Bathurst at Cirencester Park or Lord Bessborough at Stansted, Sussex, into a forest with rides and walks rather than open grounds streaked with woods and coppices. His improving hand was everywhere: thinning neglected coppice, trimming boughs of standing trees, planting roadside trees for shade and beauty and replenishing hedgerows with timber. His overall aim was to sow nurseries of trees so as to transplant seedlings on areas cleared previously, and to extend timber on worn-out fields where none had previously existed.

Representative extracts from his bailiff's weekly letters and Evelyn's Journal throw light on these developments:[9]

1707 Thinning out Deerleap Wood.
1709 Robert Lane (the gardener) transplanted 'above 500 trees of various sorts into the hedgrows and more intended'.

FIG.7.3: *ROSA GALLICA* 'VERSICOLOR' (*ROSA MUNDI*) BY PIETER WITHOOS (1654-93)

Fig.7.4: Wotton estate plan of *circa* 1720
The original plan measures 38¾ x 28 inches
Wotton House is shown, outlined in red, at lower centre

1710	Walks near the House re-laid with lime.
1712	Replanted Pasture Wood (several hundred acres).
1714	Replanted Heathy Lane Wood (250 acres).
1719	Paid Arge (tenant) for saving trees in a hedgerow at Hoopwick Farm. He was not obliged to.
1722	Planted 300 young saplings on Garrot Hill (on high ground near Leith Hill). About 3,000 more are needed.
1724	Began planting at Noones (an arboretum near Friday Street Mill Pond).
1729	Visited West End Farm, below Leith Hill, and saw in a coppice not above two acres, 1,000 tellers.
1730	The 'account of timber' above the hill (i.e on the sandy flanks of Leith Hill) showed, towards the end of the afforestation period, oaks, 6,153; beech, 2,907; ash, 489; elm, 114; oak tellers, 3,000. Below the hill the figures were 4,972, 841, 849 and 320 respectively and tellers numbered more than 20,000. Additionally fir plantations on Leith Hill extended over 19 acres.

With the aid of an estate plan of *circa* 1720[10] [Fig.7.4] the extent of the tree plantations and their distribution can be gauged for the part of the estate nearer the house. Particularly notable are the numerous fields given over to trees along the sides of the River Tillingbourne downstream of the mill-hamlet of Friday Street and on the branch of the same stream in the Broadmoor valley across the intervening ridge. These two valleys and the intervening ridge were later to be the main arena of Evelyn's landscaping activities. John Rocque's *Map of Surrey* on a scale of two inches to one mile (1768) is towards the end of Sir John Evelyn's planting career and gives a good impression of the wide bounds of afforestation under Sir John. When William Upcott was preparing John Evelyn's diary for publication he rejected an engraving of woods near Wotton House on the grounds that it appeared to show coppice instead of the magnificent grove of mature trees.

Sir John Evelyn appears to have followed in his landscaping Stephen Switzer's injunction 'to throw the garden open to all views to the unbounded felicities of distant prospect and the expanse of nature itself', reflecting the intense susceptibility of his generation to grandiose romantic nature.[11] He might well have been mindful of estates such as Wotton when recommending in 1718 that the garden designer 'should make the best of all hills and dales, of all corn fields and tufts of trees and

FIG.7.5: FRONTISPIECE TO STEPHEN SWITZER'S *ICHNOGRAPHIA RUSTICA*

wherever they strike the eye 'tis there he should create more ... Some of our seats and that not far from London have a nobleness of prospect much exceeding anything that France or Italy can boast of ...'

The plan and map afford some evidence as to how the re-afforestation reflected the principal precepts of Stephen Switzer's *Ichnographia Rustica* (1718 and 1742) [Fig.7.5]: '... the pleasures of a country life cannot possibly be contained within the narrow limits of the greatest garden; woods, fields and distant enclosures should have the care of the Industrious planter'. It is also noteworthy that Sir John also followed the Switzer recommendation of raising a nursery of trees to stock and plant the 'exterior parts' of the estate. Unfortunately no visual evidence has survived of the care Evelyn took in varying the shapes and colours of his trees and the skill by which he massed them. His interest in the chestnut [Fig.7.6], until the 1730s rarely planted in hedgerows in England, is an indication of the variety and new beauty he aimed at, and repeatedly his tree collection was enhanced by exotic species from America and the West Indies. His admiration of the noble view of St. Paul's some twenty-four miles away determined his choice of the steeply-rising, west-facing slopes above the mill-hamlet of Friday Street [Fig.7.7] as one of the major sites of an arboretum through which he would have devised a vista to the cathedral. Seats and enclosures placed round ancient

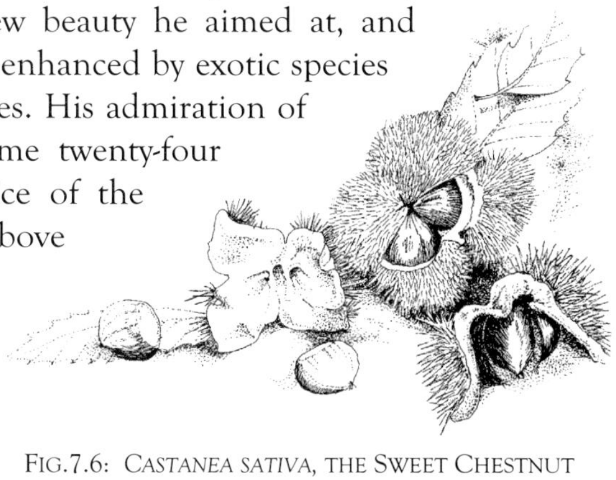

FIG.7.6: *CASTANEA SATIVA*, THE SWEET CHESTNUT
DRAWING BY CHRIS HOWKINS

beeches were particularly valued in the family because the diarist had successfully spared them from his brother's axe.

Of particular interest in the correspondence between Sir John and his bailiff are numerous references to the supply of trees from Wotton nurseries to large parks and gardens (including royal parks). Evidently Sir John's nurseries became so celebrated that he supplied many thousands of trees to neighbouring estates which were adopting his own planting

FIG.7.7: *FRIDAY STREET,* BY SUTTON PALMER (1906)

example. Wotton should thus be seen as one of the mainsprings of a landscaping movement which eventually modified much of Surrey and the entire rural character of England. As early as 1715 Evelyn had supplied 3,400 beech from his own raising to John Parker of Neden.[12] By 1723 he was growing both beech and fir from seed and had supplied large quantities to Sir Edward Nicholas of West Horsley. Evelyn had good cause to boast to Nicholas that 'for fir seed in my opinion you need go no farther than the trees of my own raising [these] being to the best of my observation as flourishing as any I have seen in the country'.[13] King George I's estates were also in receipt of trees as was William Moore's Polesden Lacey where he was carrying out extensive alterations between

FIG.7.8: CLAREMONT PARK, PAINTED FROM NATURE AND ETCHED BY HARRIOT GOULDSMITH, PUBLISHED 1819

1735 and 1748. King George II's estates were also supplied from Wotton. In some instances gifts of trees and plants were exchanged. Thus, for example, Revell of Fetcham sent Evelyn a gift of Spanish acorns which was reciprocated by a gift of Scotch fir. The longest lasting connection between Wotton and another 'improving estate' was with the Duke of Newcastle's Claremont at Esher [Fig.7.8]. Between 1731, when an annual despatch of tree plants to the Duke was already an established custom, and 1746 there was scarcely a year when Claremont was not supplied with trees by Evelyn. In 1731, for instance, the Duke of Newcastle's gardener was supplied with 1,000 chestnuts and on other occasions with two hundred or more beech and oak. These annual consignments of trees to Claremont resulted in a letter from the Duchess of Newcastle to Lady Evelyn which is a model of patrician insouciance and brevity:

March 12, 1739/40

Dear Madam,

I am really ashamed of what I'm going to ask but the Duke of Newcastle has the courage enough to beg Sir John Evelyn would be so good to let him have ten or a dozen more pretty large Scotch firs from Wotton and if he comply's with this request, the Duke of Newcastle begs he would send his orders to the gardener, because this is the time they should be planted. I think when people are unreasonable, the less they say for themselves the better, therefore I wish only to ask your pardon, and spare you, dear Madam, much

From your faithful servant,

Newcastle[14]

A keynote of Sir John's naturalistic forestry and gardening was his conscious attempt to harmonise design with local scenery. In this he was probably the first to make the Surrey hills a leading motif. Leith Hill, which lay on his estate, had a profound influence on his plantings. Its commanding view of the South Downs and the intervening Weald was becoming better known with the new passion for prospects of 'extensive and variegated views' [Fig.7.9]. The 'hill' or 'prospect' poem had become well established since Sir John Denham's 'Cooper's Hill' and Claude's paintings were the most popular prospects in landscape scenery. As early as 1658 William Sanderson in his *Graphice* had explained how to draw a prospect 'from the life' and by his definition 'landskip' became synony-

FIG.7.9: LEITH HILL, VIEW FROM THE TOP OF THE TOWER LOOKING SOUTH-EAST OVER THE WEALD TO THE SOUTH DOWNS

mous with prospect. Positive aesthetic sensibility to prospect also runs through the contemporary theoretical contributions to natural philosophy. Shaftesbury in 1709 extravagantly praised the awe-inspiring aspects of nature and Addison in 1712 told his *Spectator* readers that 'a beautiful prospect delights the Soul'. The diarist, who had already sought out beautiful prospects in France and Italy in the 1640s, remarked in Edmund Gibson's edition of Camden's *Britannia* (1695) of the view from Leith Hill that 'the like, I think, is not to be found in any part of England, or perhaps Europe besides; and the reason why it is not more observed is partly it being quite off the road ...' The view was certainly one of the finest and most extensive beheld. St. Paul's was then visible to the north, though now this view is sadly obscured by recently-grown trees, and the great expanse of the Wealds of Sussex and Kent was spread to the south. In 1717 the critic John Dennis regarded the view from Leith Hill superior to any he had seen in the Roman Campagna or elsewhere in Italy.[15] It appears evident that by Sir John's day a liking for 'an extensive and variegated view' had become a dominant characteristic of English taste in landscape.

Sir John Evelyn was excited by the sweeps of prospect afforded by his own estate from other properties. His visit to Syon House in Isleworth in 1722, for example, was made especially memorable by the view from the gallery 'which leads in a delicate prospect into Surrey as far as Leith Hill which I could plainly discern'.[16] In the same year, at Wimble-

FIG.7.10: POSTCARD SHOWING LEITH HILL TOWER BUILT IN 1766. IN 1864, MR W. EVELYN ADDED AN UPPER ROOM AND BATTLEMENTS WHICH RAISED THE TOWER TO 1,000 FEET ABOVE SEA LEVEL.

don House, his journal entry is confined to his admiration for the 'great view' from the terrace towards Box Hill and Leith Hill [Fig.7.10].[17] An earlier instance of his love of scanning extensive panoramas was in 1720 when he observed of a tenant's farmhouse that it had 'one of the largest and finest prospects I ever saw, being situated on a hill which looks down upon the town of Dorking and as far as the South Downs and twenty miles off into Kent'.[18] In 1728 Newlands Corner was visited for the sake of its views[19] and in 1729 the fine prospect of Richmond Hill was appreciated from the Duke of Argyll's villa at Petersham.[20] In 1748, however, Evelyn strikes a new note when contemplating a view near Guildford. This was the year that the young William Gilpin published, anonymously, his aesthetic *Dialogue ... upon the Gardens at Stow*, introducing his new concept of the 'picturesque', which he would later define as 'that kind of beauty which would look well in a picture'. Acting the part of a very early picturesque tourist, Evelyn wrote that 'I was pleased with the prospects St. Catherine's Hill afforded me. If I have a right notion of a landscape it wants nothing but hanging rocks and that in some measure is remedied by two large chalk pits'.[21] Ruggedness, he had learned from Gilpin, was an essential component of a landscape picture. The diarist had spoken of the wooded glens and rocky outcrops of his birthplace as 'horridly agreeable' but this was pleasure derived from the 'romantick' fancy of the age.[22] (See Appendix)

The influence of prospect upon Evelyn is very apparent in his layout of outlying parts of his estate where the controlling features were the walks of Scots pine aimed at the distant but ever-beckoning horizon of the South Downs. These were connected with conjoined plantations of deciduous trees and the diarist's formal garden about Wotton House itself with alleys lined with elm and chestnut. The outcome was not so much geometrical as a pattern of 'naturalized' avenues on irregular courses related to the natural beauties of the site. Evelyn's tree-lined walks and rides possess the landscape without dominating it. They are not aligned on the highest point of Leith Hill (on which Richard Hull of Leith Hill Place later built the famous tower) but to a focal point slightly north-west in order to converge on a horse road serving the estate's claylands 'below the hill' itself. Another conspicuous feature was the winding rides on the peripheries of his woods which formed a continuous circuit and brought one back to one's starting point, again

FIG.7.11: THE SPRUCE FIR TREE FROM *SILVA* (1801 EDITION)

In *Sylva* (1679 edition) Evelyn writes of Fir trees: 'They make our best Mast, Sheathing, Scaffold-poles, &c. heretofore the whole vessel: It is pretty (saith Pliny) to consider, that those Trees, which are so much sought after for Shipping, should most delight in the highest of Mountains, as if it fled from the Sea on purpose, and were afraid to descend into the Waters. With Fir we likewise make all intestine works, as Wainscot, Floors, Pales, Balks, Laths, Boxes, Bellies for all Musical Instruments ... It is also excellent for Beams, and other Timber-work in Houses, being both light, and exceedingly strong.

designed for utility as well as recreation. Evelyn's walks on the outlying parts of his estate were lined by platoons (Fr. *peloton*), square clumps of ten to twelve firs planted within small cattle-proof enclosures and sited at uniform intervals. Within each enclosure the firs were planted in rows analogous to the 'hollow square' disposed by a body of musketeers. When Arthur Young observed them on a visit to Wotton in 1793[23] he attributed them wrongly to 'the author of the *Sylva*'. In Sir John's journal and letters from staff there is ample evidence of the gradual construction of the original enclosures to Leith Hill. Thus in October 1727 he records: 'Cut out a circle on the common of 200-foot diameter, for my great walk of clumps of firs to be continued'. This site can be identified with that on the southern edge of Abinger Common where Rocque shows interconnecting vistas. These 'new walks on the common' were still being laid out in November 1736 when the bailiff wrote: 'We keep on amaking Pounds in the Common',[24] and 'fir planting on the common' was still being carried out as late as 1745, indicating that Evelyn was employing the minimum of labour on his improvements. Objects to embellish views from the walks were added sparingly. In 1738 gardeners were preparing timbers for an ornamental building to be set up opposite the Great Walk, but there is no further reference to this eye-catcher. The summit of Leith Hill was planted with trees in 1730 and several species of fir [Fig.7.11] were subsequently planted there. Evelyn went on to select keynote features in his landscape – bluffs, ridges, summits, ponds, river banks and similar places of agreeable character and accentuating their beauty with trees and, occasionally, buildings. These appear to have been the prototype of landscape increasingly laid out on the common lands in the valley of the Tillingbourne between Dorking and Guildford.

When his layout of walks and rides was largely completed, Evelyn turned his attention to his final phase of landscaping, which was to introduce winding rivulets and cascades. His earliest work was between *circa* 1736 and 1740 which makes it one of the earliest in the country – but rather later than the 'digging of a river' at Castle Howard in 1732 and perhaps Kent's similar experiment at Lord Burlington's Chiswick Villa. The decisive influence on Evelyn's water features appears to have been the works for the Earl of Carlisle at Castle Howard, for Evelyn made two visits there in 1732, when he observed the serpentine river under construction, and again in 1738 when his 'own river' was being serpentined [Fig.7.12]. Basically his 'New River' was a reconstructed course of the River Tillingbourne which was made to flow over an artificially-constructed waterfall below Friday Street and a series of low falls between fish ponds. At the same time the branch of the stream flowing from the Broadmoor valley was also serpentined. The work involved enlarging the bed of the river and cutting its course in a more irregular and winding manner through sheets of gravel which was under way before November 1736 when Evelyn wrote to his bailiff: 'I forgot in my letter to set down carrying on the serpentine river in the meadow ...'[25] Progress was slow. He again wrote from London in August 1737: 'Since there is no want of men, I hope you have employed as many in the meadow these fine days'.[26] In May 1738 stone quarried from Westgate Heath was being laid down for the waterfalls, by which time Evelyn's neighbour, Theodore Jacobsen, the architect of the Foundling Hospital in London, was constructing his cascade and waterfalls at his property of Lonesome Lodge [Figs.7.13 & 14]. It is not clear which of these two influenced the other with fastidious stone and brickwork.

FIG.7.12: SIR JOHN EVELYN'S SERPENTINING IN WOTTON MEADS

By 1740 Evelyn's most active period as landscapist was over, although he was left with nearly a quarter of a century in which to enjoy the fruits of his work. He had not failed to carry out the injunction of his grandfather to 'Restore the name of Wotton, otherwise in danger to be lost or forgotten'. His woods and rivers and some of his trees remain but none of his walks which today are traceable only on old maps and by the existence on the ground of relics of the 'fir pounds', the purpose of which has been forgotten locally. Thus the Wotton landscape has come to bear two quite different states of the Evelyn imagination. The diarist's strong sensibility in the mid-seventeenth century to Italian landscapes in the Renaissance tradition was to be complemented by his grandson's

FIG.7.13: LONESOME LODGE: THE WATERFALL

receptivity to Wealden nature itself in one of the most dramatically contrived examples of the early eighteenth century.

Altogether Sir John's landscaping reveals a deep appreciation of natural scenery, especially that of Leith Hill, the presiding genius of the place. His accepted principles of landscape and his manner of working was particularly appropriate to a descendant of the author of *Sylva* which marked the beginning of the notion that a countryside could be beautiful as well as productive. 'Consult the Genius of the Place in all', the famous maxim of Alexander Pope (who expressed a wish to visit Wotton) also reflected the 'beautiful simplicity', 'good sense' and 'easy politeness' that was the mark of the diarist. It seems evident that Sir John Evelyn enjoyed Wotton on his retirement from political life in London and that it was commensurate with his finances which left him in a comfortable but far from wealthy position. Victorians were powerfully attracted to the Evelyn woods. Vicat Cole, RA, painted them beautifully in the 1860s.[27] As late as 1869 Matthew Arnold was receptive to the dream-like quality of Wotton which Sir John had striven to make[28] and Richard Redgrave, RA, was so inspired by its woods that he spent thirty-six successive summers painting at Abinger towards the end of the nineteenth century.[29] [Fig.7.15]

FIG.7.14: LONESOME LODGE: THE THATCHED FISHING-HUT BUILT TO RESEMBLE A RUSTIC GROTTO

FIG.7.15: *PARKHURST WOODS, ABINGER, SURREY* BY RICHARD REDGRAVE (1804-88).

REFERENCES

The author wishes to thank Frances Harris for her help in changing Christ Church Oxford references to the present British Library sequence.

1 For general background see P. Brandon, 'Land, technology and water management in the Tillingbourne Valley, Surrey, 1560-1760', *Southern History*, Vol. 6 (1984), pp. 75-103; *The Story of the Tillingbourne*, (Shere, Surrey, 1984 and later editions).
2 John Evelyn, *Silva: or a Discourse of Forest-Trees* (1706 edition), p. 581.
3 W.G. Hiscock, *John Evelyn and his Family Circle*, (London, 1955), pp. 201-3, 236-7.
4 British Library (hereafter BL) Add. MS 78298.
5 John Bowle, *John Evelyn and his World* (London, 1981), p. 214. See also Hiscock, op.cit., p. 235.
6 John Evelyn, *Memoires for my Grand-son*, ed. Geoffrey Keynes (Oxford, 1926), pp. 16, 17, 19, 20, 28-9, 30-1.
7 Sir John Evelyn became a Baronet in 1713. He was admitted a Fellow of the Royal Society and a Fellow of the Society of Antiquaries of London, eventually becoming a member of its Council.
8 BL Add. MS 78482.
9 BL Add. MSS 78482, 3, 4, 5, 6, & 7.
10 Surrey History Centre, Woking, 329/15.
11 Stephen Switzer, *Ichnographia Rustica*, vol. 1, p. xxxv

12 BL Add. MS 78484.
13 BL Egerton MS 2540, f. 233.
14 BL Add. MS 78470.
15 J. Dennis, *Original Letters* (1721), pp. 31-4. There was a similar response from John Burton in 1752. BL Add. MS 11571.
16 BL Add. MS 78514B, f. 67.
17 BL Add. MS 78514B, f. 63.
18 BL Add. MS 78514E, f. 8.
19 BL Add. MS 78514B, f. 11.
20 BL Add. MS 78514B, f. 20.
21 BL Add. MS 78528.
22 John Aubrey, *The Natural History and Antiquities of the County of Surrey* (1718-19).
23 Arthur Young, *Annals of Agriculture*, vol. 22 (1794), pp. 195-6, 199-200. The 'noble plantations' of Wotton were also noted.
24 BL Add. MS 78490.
25 BL Add. MS 78490.
26 BL Add. Ms. 78490.
27 Robert Chignell, *The Life and Paintings of Vicat Cole, RA*, vol.1 (London, 1898).
28 Matthew Arnold, *Letters*, vol.2, (London, 1895), pp. 201 and *passim*.
29 This fact is commemorated by a plaque in the parish church of Abinger [Fig.7.16].

FIG.7.16: PLAQUE TO RICHARD REDGRAVE IN ABINGER CHURCH

PETER BRANDON is an historical geographer with many published works to his credit. Formerly the Head of the Department of Geography at the University of North London, he was a founder member of the Sussex Downs Conservation Board, editor of *Sussex Archaeological Collections,* and served as Chairman of the Sussex branch of the CPRE for thirteen years. His publications include *The South Downs* (1998), *The Kent and Sussex Weald* (2003) and *The North Downs* (2005). The Surrey Hills, where he walked and cycled in his youth, were his first love. Undergraduate classes at the Juniper Hall Field Centre near Dorking started his lifelong passion for the landscape of South-East England.

FIG.8.1: SOUTHERN RAILWAY POSTER (1926)
'THE LONDONER'S HIGHLANDS – SURREY'

8

JOHN EVELYN AND THE LANDSCAPE GARDEN: 'TO HIM IT IS OWING THAT GARDENING CAN SPEAK PROPER ENGLISH'

MAVIS BATEY

Landscape gardening is widely appreciated internationally as a unique contribution to the visual arts and typically English in its idealisation of our countryside in a union of beauty and use. Stephen Switzer, writing in 1718, was convinced that Evelyn had played a significant part in establishing this landscape tradition of rural gardening, which was not called landscape gardening until the end of the eighteenth century. Although over the centuries much attention had been paid to horticulture in numerous publications, the history of garden design, unlike that of art and architecture, was only brought together as a cohesive subject when the Garden History Society was founded in 1965 and published its journal *Garden History,* edited by Dr Christopher Thacker in 1972. The following year under the auspices of the Courtauld Institute the origins of the landscape garden were debated and a paper by Denis Lambin of the Sorbonne published in the journal.[1]

The French view considering how and why England rejected the continental baroque grand manner in favour of rural landscape gardening was particularly perceptive as Professor Lambin saw Evelyn's contribution to the art of gardening as the result of agricultural tradition traced back to ancient Rome, while in France he felt that 'these agricultural traditions of garden art have, in some way, – at least to my mind – been almost totally obscured by André Le Nôtre (1613-1700) who transformed vast stretches of the countryside into those architectural or theatrical scenes that have been called *jardins à la française'* [Fig.8.2]. On the other hand, his younger contemporary by seven years, John Evelyn (1620-1706), following the agricultural traditions I am going to evoke, and never swerving from them, promoted vast transformations of the English countryside and (in Switzer's words) 'to him it is owing that Gardening can speak proper English'.

FIG.8.2: EVELYN DID NOT SEE THE FINAL PHASE OF THE GRAND MANNER OF LE NÔTRE AT VERSAILLES BUT ALREADY HE DESPISED WHAT HE SAW AS 'UNIVERSALL LUXURY' AND ADVOCATED 'GROVES AND WALKS AND OTHER INNOCENT AMENITYS'

For his appraisal of Evelyn's vital role Denis Lambin consulted his internationally acclaimed *Sylva* and entries in his diary referring to French gardens he approved of, such as Cardinal de Richelieu's country house near Paris, which he visited in 1644, where, in addition to magnificent garden features, he felt that 'what is most admirable, is the vast enclosure, and variety of ground, containing vineyards, cornfields, groves (whereof one of perennial greens) and walks of vast lengths'.[2] Denis Lambin also studied Evelyn's translations of French gardening books, such as *Le Jardinier François* by Nicolas de Bonnefons and his quotes and borrowings from them and the Latin agronomists; in particular, he noted Evelyn's address to the readers of *Sylva* where he recommends Columella's works 'to the serious perusal of our Gentry'. In *De Re Rustica* Book IX Columella had suggested that 'a park with trees and game should be thus sited as never to stand too far from the villa and the master's eye both for pleasure and profit'.

Evelyn's diary entry for 16 October 1671 records that he induced Lord

Arlington, who was building Euston Hall in Suffolk, to bring his park 'so near as to comprehend his house within it' [Fig.8.3]. Evelyn's son and heir, John the Younger, had also travelled and was aware of French agricultural traditions; he translated Rene Rapin's poem 'Of Gardens' in 1673 in which the poet alludes to pre-Le Nôtre country seats with happy combinations of gardens, parks and cornfields where the landowner could

> to cool Streams, to aged groves retire
> and th' unmix'd pleasures of the field desire';

it was dedicated to his father's friend Lord Arlington of Euston.

If Professor Lambin had had the advantage of Evelyn's published *Elysium Britannicum or The Royal Gardens* he would have found much additional material to support the idea that the roots of the English natural garden were embedded in long standing rural traditions. When the elderly Evelyn retired to Wotton he told his intimate friend Samuel Pepys that he was spending his days in the fields engaged on rural matters 'in the manner which Cicero reckoned the best occupation for a senator in old age'.[3] Although Evelyn in his diary and comprehensive survey of gardens and gardening planned for his 'Elysium Britannicum' described many of the 'pompous' features he had seen in gardens in France and

FIG.8.3: EUSTON HALL, *CIRCA* 1710, BY THOMAS WYCK, WITH A STAG HUNT IN PROGRESS

Italy and, moreover, had himself designed an Italianate garden for Albury and given advice to his brother on Italianate features at Wotton, he made it clear that his ideal Elysium extended beyond the confines of the garden into the surrounding countryside with due consideration of the special nature of the place.

We do not know in what direction trend-setting royal gardening might have taken the English landscape in the mid-seventeenth century if there had been no Civil War and Commonwealth with its condemnation of luxury and strong interest in husbandry. Evelyn had left Oxford just before the Civil War broke out, but in 1654 he met up with Dr Wilkins and his Oxford Experimental Philosophy Club, which included Christopher Wren and Robert Boyle, who became Evelyn's mentor in empirical scientific thinking. His famous dictum was 'Experiment is the interrogation of Nature'.

FIG.8.4: WADHAM COLLEGE GARDEN AS DEPICTED BY LOGGAN IN *OXONIA ILLUSTRATA* (1675). THE COLLEGE THAT WAS AT THE CENTRE OF EXPERIMENTAL SCIENCE IN THE COMMONWEALTH EXTENDED ITS ENQUIRIES TO THE GARDEN WITH ARTIFICIAL RAINBOWS, TRANSPARENT BEEHIVES DESIGNED BY WREN AND A SPEAKING STATUE

Oxford had been spared by civilised moderation on both sides in the Civil War. Fairfax, who had summoned the garrison city to surrender, placed a guard on the Bodleian Library. Dr Wilkins, who was married to Cromwell's sister, was intruded as Warden of Wadham, where Wren whose royalist uncle was in the Tower, was studying. With such divided loyalties, the latitudinarian Warden refused to allow religious or political disputes in the college and encouraged studies in Baconian natural philosophy, which sought to banish myth and superstition and was equally acceptable to both sides.[4]

The 'universally curious' Warden's garden at Wadham [Fig.8.4], laid out in 1651, captivated Evelyn; it had rainbows made by water forced through narrow fissures and a speaking statue on a mound; this was not a joke feature but had pipes through to the Warden's study where he worked on speech pathology, hoping to teach deaf mutes. Evelyn recorded in his diary entry for 13 July 1654 that he was presented with one of Wren's transparent beehives adorned with dials and vanes for Sayes Court, where he had his own 'elaboratory' [Fig.8.5].

FIG.8.5: EVELYN'S DRAWING OF WREN'S TRANSPARENT BEEHIVE FROM 'ELYSIUM BRITANNICUM'

In an address to Lord Somers, Evelyn stressed the importance of scientific enquiries for enlightenment. Orthodoxy and scholasticism had for centuries impeded progress in the universities and, looking back to those days, Evelyn deplored the way their disputations had led to 'fantasies and fruitless Speculations' and he commended the present experimentalism which promoted 'solid and useful knowledge by the Investigation of Causes, principles, Energies, powers and Effects of Bodies and things visible and to improve them for the Good and benefit of Mankind'.[5]

Evelyn had learned a great deal from Oxford empiricism and he would meet up with many of the Wadham set again when the Royal Society was founded with Dr Wilkins as Chairman; these were the brilliant men whom Thomas Sprat in his *History of the Royal Society* said had in 'fanatical times' been 'provided against the next Age'. Evelyn was always an active member of the Royal Society and brought gardening into its transactions. In 1662 he gave a 'Discourse Concerning Forest Trees', which was published two years later as *Sylva*; it greatly influenced landowners, now eager to improve their estates after the depletion of timber in Commonwealth days. It was the Royal Society's first publication.

Evelyn had left Oxford without a degree in 1640 but was one of the first to receive a well-earned doctorate in Wren's new Sheldonian theatre

FIG.8.6: WREN'S SHELDONIAN THEATRE, OXFORD, FROM *PHARMACEUTICE RATIONALIS* BY THOMAS WILLIS (1674)

[Fig.8.6]. Wren had already shown Evelyn the model of it in 1664 'not disdaining his advice'. It was with great pride Evelyn wrote in his diary for 9 July 1669 about the celebration of the 'Encenia in the New Theater so magnificently built by the munificence of Dr Gilbert Sheldon'. The theatre's baroque ceiling depicting Academical Learning triumphing over the dark forces of evil said it all about Oxford's joy in the Restoration. Evelyn had persuaded Lord Arundel's heir, Henry Howard, to present the Arundelian antique inscriptions to the University to be installed on the walls of the theatre precinct. He was disgusted to find that people had already begun to scratch their names on them before he left and advised the Chancellor to have prickly holly planted in front of them.

Evelyn had found a role for himself in the Interregnum by bringing gardening into the scientific movement with the help of Robert Boyle, whose laboratory in the High Street was right by the Oxford Physic Garden [Fig.8.8] where the keeper Jacob Bobart was busy cultivating, propagating and cataloguing its wide range of plants. Evelyn would acquire some of them for his own garden at Sayes

FIG.8.7: GARDENER AT WORK FROM JOHANNIS URSINI'S *ARBORETUM BIBLICUM* (NUREMBURG, 1699)

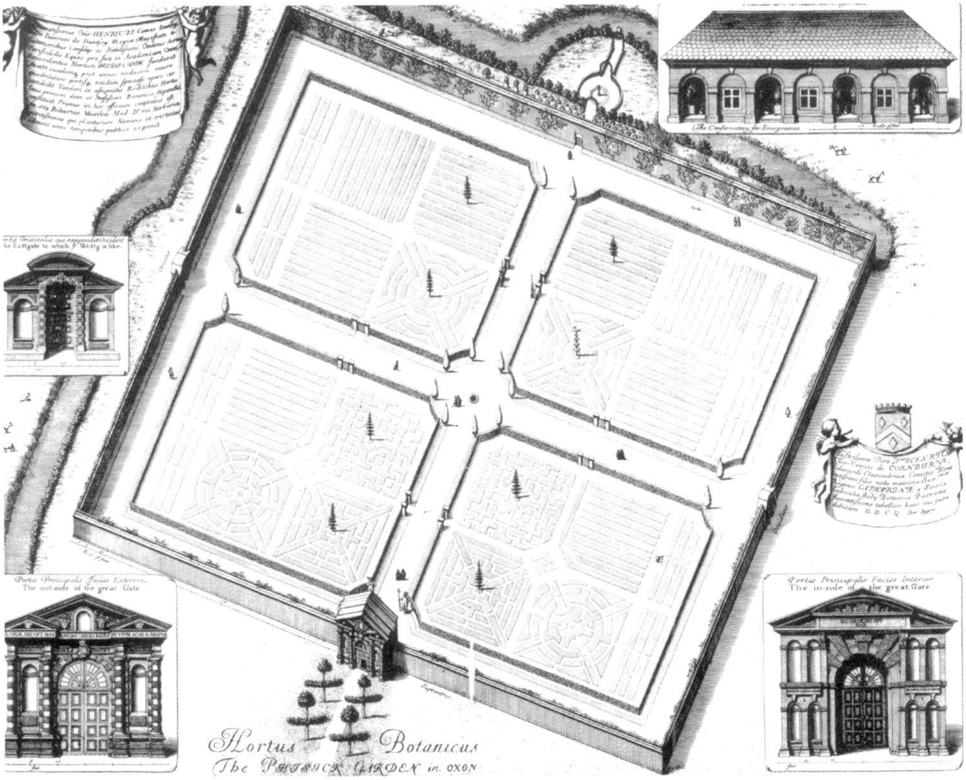

FIG.8.8: THE OXFORD PHYSIC GARDEN PLAN IN LOGGAN'S *OXONIA ILLUSTRATA* (1675) SHOWING THE EARLY GREENHOUSE IN TOP RIGHT-HAND CORNER

Court. Boyle believed that the best people to consult about good husbandry were gardeners [Fig.8.7] and ploughmen; he worked with Bobart on what was best for mulching, composting and fertilising. Both Boyle and Evelyn were interested in the conservation of plants and it is not surprising that Oxford had one of the first greenhouses in the country. Robert Sharrock of New College published, in 1659, *The History of the Propagation and Improvement of vegetables by the Concurrence of Arte and Nature, written according to Observations made from Experience and Practice,* which was dedicated to the great master of 'vegetative philosophy' Robert Boyle.

All this time Evelyn had been working on his great compendium of gardening the 'Elysium Britannicum' and in 1659, with the end of the Interregnum in sight, he made a synopsis of it for circulation. The reaction of Oxford was particularly important and it was seen by

Sharrock and the Bobarts, father and son, together with other experts, doctors and Heads of Houses; it was well received and Evelyn was encouraged to publish it. Perhaps the dons had had enough of puritan good husbandry as they seemed anxious to hear about the 'pompous beauties' in gardens he had seen in Italy.[6]

Evelyn, an ardent royalist, was on very friendly terms with Samuel Hartlib and his circle of puritan intellectuals. The advancement of knowledge in husbandry was a shared Baconian commitment although Evelyn did not go along with their radical ideas or belief in 'Adam and Eden' gardening which would hasten the Millennium.[7] Samuel Hartlib and John Beale were as anxious as the Oxford dons for Evelyn to publish his *magnum opus*, the 'Elysium Britannicum', although they were not happy with the pagan Elysium imagery. Evelyn, however, had no doubts about the title ELYSIUM, give or take God's handiwork in Eden and, ironically, it was John Beale's own garden that he would describe as his ideal site, 'proper for a most accomplished Elysium', blessed with what its owner called 'luckiness of situation' in his native Herefordshire. Hartlib remarked in 1660 that 'when Mr Evelyn's Book concerning Gardens will be finished, he will like it far better than his Garden near London'.[8] It would be another thirty years before Evelyn, then living close to the docks at Deptford, would become heir to Wotton with its own 'luckiness of situation' and opportunity for creating an Elysium in the Surrey countryside he loved.

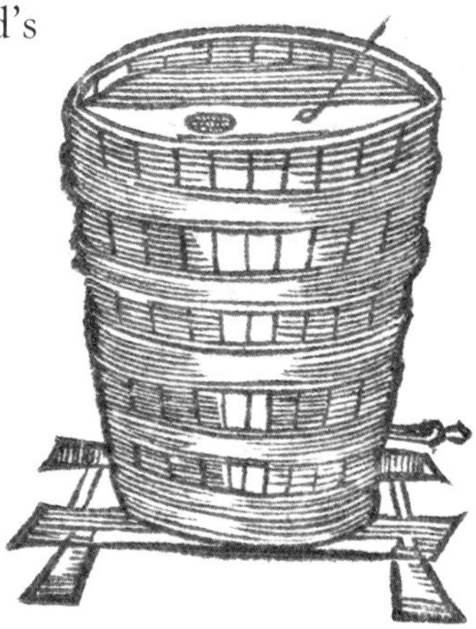

FIG.8.9: CIDER BARREL FROM 'POMONA' APPENDIX TO *SYLVA*

John Beale's garden was for Evelyn a 'sweete and natural garden' in a rich countryside contrived 'to Agree with the nature of the place' and make best use of the land, which Beale saw as 'antique gardening'. In the midst of orchard country he would contribute to Evelyn's 'Pomona' appendix to *Sylva* [Fig.8.9]. Although Evelyn described Backbury in detail in the 'Elysium Britannicum' and assured his readers that it was 'no phantasticall Utopia but a reall place', the information was second hand as he never saw it himself.[9] However, Peter Goodchild in his garden history hat did go and find it for him and made a schematic plan of the landscape for the Garden History Society's journal[10] which would certainly have helped Evelyn; it shows how the garden extended into the countryside in a way which heralded Switzer's 'rural and extensive gardening' and one which would take advantage of prospect even in wild places. The appreciation of the long view, like so much else in the perception of landscape, was literary in origin.

The most influential prospect poem was inspired by a Surrey scene. John Denham's 'Cooper's Hill' overlooking Runnymede, written in 1642, started the genre for topographical poetry. Milton who also admired a 'goodly prospect' had already taken a long view of the rural landscape in his 'L'Allegro':

> Straight mine eye hath caught new pleasures
> While the landscape round it measures.

Already in 1621 Robert Burton had said that 'to run up a steep hill sometimes, or sit in a shady seat, must needs be a delectable recreation' and a cure for melancholy.[11] By 1650 a seat had appeared on Richmond Hill [Fig.8.10], which must be the most painted prospect in England, especially after James Thomson's eulogy of it in 'The Seasons'. Landscape or landskip was a painterly term which then meant an ideal construction of a region as Henry Peacham defined it 'expressing of the land by hills, woodes, castles, seas, valleys, ruines, hanging rocks, Citties, Townes ...

FIG.8.10: *THE TERRACE AND VIEW FROM RICHMOND HILL, CIRCA* 1710-20, BY LEONARD KNYFF.
'HEAVENS! WHAT A GOODLY PROSPECT SPREADS AROUND'.
(JAMES THOMSON, 'THE SEASONS', 1727)

within one horizon.'[12] In his *The Compleat Gentleman* he had encouraged landscape sketching as a worthwhile pastime: Evelyn himself had indulged in it at an early age. Henry Peacham in the 1620s had in fact chronicled the rise of the virtuoso, an amateur, who combined art, antiquities, literature and natural philosophy in a way which characterized what the schoolboy John Evelyn was about to become, par excellence.

Lucretius in *De Rerum Natura,* which Evelyn translated in 1656, spoke of 'the surprising artifice of some various Scene, curious Landskip, or delicious prospect'. Evelyn had sought out delicious prospects on his continental tours and his friend Edmund Waller who travelled with him felt strongly that 'the prospect thought and contemplation gives'. There was a great increase in poems of rural celebration and 'green thoughts' in gardens in the Civil War.[13] Evelyn's other poet friend and associate, Abraham Cowley, who lived at Chertsey, dedicated his poem 'The Garden' to him as Evelyn did his *Kalendarium Hortense* to the poet. Evelyn wrote poems himself which Pepys said he loved to read to his friends. Groves were more than garden features for him; they were as sacred as they were to Virgil and the 'divine Petrach'. As he tells us in *Sylva*; 'In the walks and shades of trees the noblest raptures have been conceived and poets have composed verses which animated men to heroic and glorious actions'.[14]

By 1660 Evelyn had written to Sir Thomas Browne concerning his projected 'Elysium Britannicum' for all who 'delight in gardens'. Turning his back on formal gardening he explained that 'our drift is a noble, princely, and universall Elysium, capable of all the amoenities that can naturally be introduced into gardens of Pleasure ... We will endeavour to show how the aire and the genious of gardens operate upon humane spirits towards virtue and sanctitie'. The first entry in *The Genius of the Place: The English Landscape Garden (1620-1820),* published in 1975, edited by John Dixon Hunt and Peter Willis, after due reference to Francis Bacon and Sir Henry Wotton as prophetic voices for the English landscape garden, is Evelyn's letter to Sir Thomas Browne.[15] Sir Roy Strong ends his book *The Renaissance Garden in England* in 1979 with Evelyn's letter which he says opened a new chapter away from formal gardening. Evelyn's ideal Elysium anticipated natural gardening when he said that 'there is nothing less taking than an affected Uniformity in great and noble gardens'. Art, he thought, 'though it contend with Nature', yet should 'by no means jostle it out'.[16]

In 1660 Evelyn visited Cliveden with its 'stupendous natural rock, wood and prospect' which reminded him of Frascati. 'It is a romantic object and the place altogether answers the most poetical description that can be made of solitude, precipice, prospect or whatever can contribute to a thing so very like their imaginations'.[17] So, as well as leading us into the natural landscape garden, has Evelyn also taken us into the Romantic Movement with his use of the word? Not really; not any more than Horace Walpole did in the 1740s when he said he had 'romantic inclinations' when he built Gothic Strawberry Hill [Fig.8.11]; they were

FIG.8.11: *STRAWBERRY HILL, TWICKENHAM, FROM THE SOUTH, CIRCA* 1755-9 BY JOHANN HEINRICH MÜNTZ (1717-98). OIL ON CANVAS

merely evoking Spenser and the tales of romance, which is how the word is defined as late as *Johnson's Dictionary*. Evelyn quotes *The Faeirie Queene* immediately after his description of Beale's gardens on which he wants his Elysium to be modelled, recalling Spenser's Bower of Bliss.[18] He told

Browne that he intended to write an account of 'romantic and poeticall gardens' in his book.

Romantic at this stage is closely related to imagination, which forms pictures of romantic scenes and relates them to reality as Evelyn did at Cliveden or, when writing to John Aubrey, he describes the country round Wotton as 'horridly agreeable and romantic'.[19] When he visited St Vincent's rock above the Avon gorge at Clifton in 1654, he recorded that 'there is also on the side of this horrid Alp, a very romantic seate'.[20] Horrid, as we have seen, did not mean horrible but was a poetic word implying ruggedness and a hundred years later Gilpin, visiting this spot, would find this same quality 'in a great degree picturesque'.[21] The Oxford empiricists had already investigated the faculty of sensibility in the imagination in some depth. Dr Robert Plot, the first keeper of the Ashmolean Museum, whom Evelyn frequently quotes, is rapturous about the 'noble and useful Discovery' in the anatomical research of Dr Willis of Christ Church in dissecting the brain [Fig.8.12].[22] Evelyn visited his Anatomy School in 1654. Wren had made a realistic pasteboard model of the brain which located the actual apartment for what Plot called the 'sensitive soul' as distinct from the 'rational soul'.

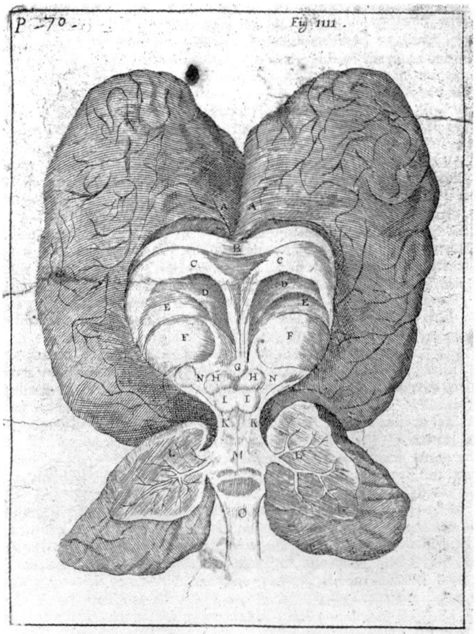

FIG.8.12: ANATOMY OF THE BRAIN FROM *THE REMAINING MEDICAL WORKS OF THAT FAMOUS AND RENOWNED PHYSICIAN DR THOMAS WILLIS*, 1681

Henry More pins it down further and, if we had any doubts, shows just what the romantic imagination signified in Evelyn's empirical world. 'As for the Imagination, there is no question but that Function is mainly exercised in the chief seat of the Soul, those purer Animal spirits in the fourth ventricle of the Brain. I speak especially of that Imagination which is most free, such as we use in romantic inventions'.[23]

Into this empirical world of demonstrating how the brain and nervous system functioned stepped John Locke of Christ Church [Fig.8.13], a

FIG.8.13: JOHN LOCKE, ENGRAVING BY S. FREEMAN AFTER SIR GODFREY KNELLER, 1697

lifelong friend of Robert Boyle, and Evelyn's 'excellent learned gentleman', a philosopher with a medical degree, who wanted to ascertain just what the brain was capable of understanding. Of Locke's *Essay Concerning Human Understanding* published in 1690, Evelyn wrote to Pepys that 'so rare and excellent a piece ought to be required reading for all intelligent readers' and urged that a medal should be made to honour him.[24]

Locke's disciple, Joseph Addison of Magdalen College [Fig.8.14], followed this by investigating what was pleasurable to the imagination and came out forcibly on the beauties of prospect and argued that the works of nature were more pleasing to the imagination than those of art; this was written in Oxford in the late 1690s but not brought to general attention until 1712 when Addison wrote his 'The Pleasures of the Imagination' essays for the popular *The Spectator* magazine.[25] By then he had had, like Evelyn, the Grand Tour experience of seeing French and Italian gardens and, as he did, particularly admired the extensive royal forest landscape of Fontainebleau.

Addison saw the vast grand new Le Nôtre gardens and thought that so much ground should not be alienated from pasturage and the plough. His answer was to plan an extensive landscape by incorporating cultivated fields and wilderness with ornamental planting into estate gardening and he asked his readers in his conversational Spectator manner – 'But why may not a whole estate be thrown into a kind of garden by frequent plantations, that may turn as much to the profit as the pleasures of the owner?'[26] This was immediately taken up

FIG.8.14: JOSEPH ADDISON (1672-1719)

by Switzer, the landscape designer in waiting, with due acknowledgement to the lead given by Virgil and Evelyn, when in 1715 he published the first volume of his *Ichnographia Rustica* to show practically how 'a man might make a pretty landskip out of his own possessions', as Addison recommended.

FIG.8.15: *A VIEW OF WOOBOURN IN SURRY, THE SEAT OF PHILIP SOUTHCOTE ESQ.* (1759)

Switzer particularly favoured a 'Farm-like Way of Gardening', which he said 'is really the truest and best way of Gardening in the World, and such as the politest and the best Genius of all Antiquity delighted in'.[27] The two most famous ornamental farms were Woburn Farm [Fig.8.15], Surrey, by Philip Southcote and The Leasowes, Worcestershire, by William Shenstone, both of whom, as their own designers, had succeeded in making 'a pretty landskip out of his own possessions'. Gainsborough's Mr and Mrs Andrews [Fig.8.16] are permanently recorded as landowners who agreed with Addison that 'Fields of corn make a pleasant Prospect'.

It was an auspicious moment for Switzer and the Brompton Park Nursery for whom he worked to encourage the peaceful pursuit of 'rural and extensive' landscape gardening. The Treaty of Utrecht in 1713 concluded the War of the Spanish Succession, France was defeated; the Glorious Revolution Settlement had been ratified by the Hanoverian Succession; the Jacobite Rebellion was being quashed and the country set on a course of prosperity. As Pope said 'Chiefs out of War' were happy to 'hang up their old trophies on their garden gate'. No longer could a man be impeached or his property seized on account of religion or politics, and the laying out of extensive grounds with long term 'frequent plantations' was a demonstration of faith in the Constitution.

Although by this time John Evelyn had left the scene, for many improving landowners his great forestry classic *Sylva,* which went through five editions, was their planting guide. Peter Brandon's researches have shown how the new heir of Wotton, Jack Evelyn, who inherited in 1706 (his father John the Younger having predeceased him) carried out his grandfather's instructions for restoring and improving the Tillingbourne Valley estate. An early Evelyn diary entry praising Surrey's gardens had received a marginal note condemning the 'later and universall luxury of the whole nation since abounding in such expenses'. By then he had probably seen William III's parterres at Hampton Court. Evelyn's

FIG.8.16: *MR & MRS ANDREWS* BY THOMAS GAINSBOROUGH, *CIRCA* 1750.
'FIELDS OF CORN MAKE A PLEASANT PROSPECT' (ADDISON)
ROBERT ANDREWS OF AUBERIES HAD MARRIED FRANCES CARTER OF NEIGHBOURING BALLINGTON HOUSE, NEAR SUDBURY, AND SO JOINED THEIR PRETTY RURAL POSSESSIONS

grandson was cautioned against any such garden luxury and instructed to invest in 'Groves and Walks and other innocent Amenitys'.

Jack, or rather Sir John Evelyn as he became in 1713, was so successful in his landscaping by 1722 that Alexander Pope, who was making his own name as an authority on landscape gardening, expressed a wish to see his improvements.[28] Pope knew him through his friendship with Simon Harcourt, whose wife was Sir John's sister. Simon died in 1720 and Sir John became the guardian of his nephew who inherited the Harcourt estates aged six, becoming the 1st Earl Harcourt. The young man also inherited all the Evelyn love of beautiful prospect and in the 1750s planned at Nuneham Courtenay [Fig.8.17], overlooking the Thames, what Walpole called 'one of the most beautiful landscapes in the world'.

Pope's sensibility to landscape had been expressed in his early pastoral poetry written in 1709 when he lived on the edge of Windsor forest:

> See what Delights in Sylvan Scenes appear
> Descending Gods have found Elysium here.[29]

FIG.8.17: VIEW OF NUNEHAM BY PAUL SANDBY (1760). SIR JOHN EVELYN WAS THE GUARDIAN OF THE IST EARL HARCOURT WHO WOULD BUILD A VILLA OVERLOOKING THE THAMES NEAR OXFORD AND CREATE 'ONE OF THE MOST BEAUTIFUL LANDSCAPES IN THE WORLD'

Addison and Pope were the first to condemn artificial gardens and to promote naturalised gardens, given that Evelyn's thoughts about Art and Nature and the desire to landscape in a way which would agree with the 'nature of the place', remained in manuscript. Addison died in 1719 and

FIG.8.18: ESHER PLACE BY W.WOOLLET. EVELYN RECOMMENDED THAT AN OWNER SHOULD BRING HIS PARK 'SO NEAR AS TO COMPREHEND HIS HOUSE WITHIN IT'. SUCH IDEAS WERE BROUGHT TO FRUITION BY WILLIAM KENT AT ESHER PLACE IN THE 1730S

it was left to Pope to instruct his 'bold associate', the painter William Kent, to work out a system. Forgetful of what went on before, Horace Walpole promoted Kent as the pioneer genius who leapt the fence and united garden and landscape.

One of the new discoveries is that Sir John Evelyn had his own nurseries and supplied trees in best *Sylva* tradition to Esher [Fig.8.18] and Claremont for Kent in the 1730s, two places that were highly praised by Walpole and Whately as the best examples of 'modern gardening'. Walpole maintained that it was in the management of water that Kent surpassed himself – 'Adieu to canals, circular basons, and cascades tumbling down marble steps ... The gentle stream was taught to serpentine seemingly at its pleasure'.[30] Sir John Evelyn certainly knew Kent by 1740 when he built a London house for him but there is no evidence that he had consulted him about managing the Tillingbourne stream. We now know that Sir John had already given his bailiff instructions about a serpentine river in the 1730s.

Addison's homely squire, Sir Roger de Coverley, had a much admired neighbour called Leonora, a lady whose 'reading has lain very much among Romances'; he delighted in her 'shady walks' by a 'little Rivulet which runs through a green meadow and was known in the family as the Purling Stream'.[31] Switzer, mindful of the romantic Leonora, would ask in the manner of Addison's Spectator 'if there were trees for Shade with little walks and purling streams, mix'd and incorporated one with the other, what cou'd be more diverting?'[32]

The word serpentine for garden design came into common parlance in 1730 when the *London Journal* announced that they were to begin digging the Serpentine in Hyde Park as her Majesty shall direct. Queen Caroline was a friend and admirer of Addison and considered herself the true patroness of landscape gardening, 'helping Nature and not losing it in art 'as she called it.[33] But haven't we heard all this before and didn't Jack Evelyn have access to it long before anyone else? All he had to do was to turn to Chapter IX of his grandfather's unpublished 'Elysium Britannicum' to read:

> Let the rivulets in gardens be narrow and shallow and let it be don as naturall as may be and so cutt as not to glide altogether in a straite line, but in frequent meanders and serpentings.

We now hear that George London, Evelyn's 'excellent Gard'ner worthy of his Royal Title', had been called in by his grandson as a consultant which adds further continuity to the Wotton story. London had discussed *Acetaria* with Evelyn and had helped to get his *The Compleat Gard'ner*, a translation of La Quintinye, into print in 1693. In the Preface Evelyn praised the London and Wise nursery at Brompton Park. Although Addison thought they erred on the side of promoting parterre gardening, topiary and moveable plants to their clients, Evelyn clearly thought that they could also provide for his type of Elysium landscape when he wrote; 'They have a very brave and noble assembly of the flowery and other trees ... and understand what best to plant the humble boscage, wilderness or taller groves with'. Who else, then, would his grandson turn to for advice? Switzer was apprenticed to London and his partner, Henry Wise, in 1709 but it is doubtful whether he was the skilled assistant they sent to Wotton to prune the fruit trees in 1715, as by then he was well on the way to making a name for himself as an independent garden designer and writer.

It was the year Switzer published the first volume of *Ichnographia Rustica* with its subtitle of the Nobleman, Gentleman, and Gardener's Recreation containing Directions for the Surveying and Distributing of a Country-Seat into Rural or Extensive gardens, by the ornamenting and Decorating of Distant Prospects, Farms, Parks, Paddocks, etc. As the new research has shown, Wotton became famous in the eighteenth century but today when Leith Hill and Friday Street [Fig.8.19] are seen as public beauty spots in the Surrey Hills Area of Outstanding Natural Beauty,[34] it is not appreciated that they were part of the Evelyn landscaping. Sir John Evelyn must have felt an obligation to pay heed to Switzer's directions when the book spoke in such glowing terms of his grandfather who, 'like another Virgil ... was appointed for the retrieving the calamities of England and re-animating the spirit of his countrymen for their planting and sowing of woods – to him it is owing that gardening can speak proper English'.[35]

FIG.8.19: FRIDAY STREET WAS A POPULAR TOURIST ATTRACTION IN THE 1920S

REFERENCES

1 Denis Lambin, 'Pleasure with profit: some aspects of landscape gardening', *Garden History*, vol. 3, no. 2 (1975), pp. 29-36.
2 E.S. de Beer (ed.), John Evelyn's *Diary*, vol. 2, pp. 109-10, 28 February 1644.
3 Peter Brandon, *The North Downs* (Chichester, 2005), p. 108.
4 Mavis Batey, *Oxford Gardens* (1982), pp. 39-40.
5 Preface to *Acetaria* (1699), published when Evelyn was Secretary of the Royal Society and Lord Somers its President.
6 John Ingram (ed.), *Elysium Britannicum* (2001). 'To encourage the finishing of Elysium Britannicum', Appendix 11.
7 For Adam and Eve gardening see *Oxford Gardens*, pp. 37-8.
8 *The Diary and Correspondence of Dr John Worthington*, vol. 1, p. 174.
9 *Elysium Britannicum*, p. 97.
10 Peter Goodchild, 'No Phantasticall Utopia, but a Reall Place. John Evelyn, John Beale and Backbury Hill', *Garden History*, vol. 19, no. 2 (1991).
11 Robert Burton, *The Anatomy of Melancholy* (1621), pt. 2, sect. 2, mem 4.
12 H.V. Ogden & M.S. Ogden, *English Taste in Landscape in the Seventeenth Century* (1955), p. 5.
13 The most famous one being Andrew Marvell's 'The Garden' with its 'green thought in a green shade'.
14 *Sylva* (1670), p. 238.
15 Geoffrey Keynes (ed.), *The Letters of Sir Thomas Browne* (1931), vol. 6, p. 305.
16 *Elysium Britannicum*, p. 96.
17 *Diary*, vol. 4, p. 153.
18 Edmund Spenser, *The Faeirie Queene*, Book 2, canto 12, stanzas 58-9.
19 John Aubrey, *The Natural History and Antiquities of the County of Surrey* (1718-19). Evelyn's contribution was written in the 1670s. See Appendix.
20 *Diary*, vol. 3, p. 103, 27 June 1654.
21 William Gilpin, *Observations on the River Wye and several parts of South Wales, etc. relative chiefly to Picturesque Beauty; made in the summer of the year 1770* (London, 1782), p. 92.
22 Robert Plot, *The Natural History of Oxfordshire* (1675), chapter IX, para. 216.
23 Henry More, *The Immortality of the Soul* (1658), p. 228.
24 Evelyn to Pepys, 26 February 1690, quoted in *Particular Friends: The Correspondence of Samuel Pepys and John Evelyn*, ed. Guy de la Bédoyère (2005), pp. 213-5.
25 The MS edition belonged to J. Dykes Campbell who published it as *Some Portions of Essays contributed to the Spectator by Mr Joseph Addison* in a limited edition in 1864; the single essay was developed into Nos. 411-421 in *The Spectator* (1712).
26 *The Spectator* 414, 25 June 1712. See Mavis Batey 'The Pleasures of the Imagination: Joseph Addison's Influence on Early Landscape Gardens', *Garden History*, vol. 33, no. 2 (2005).
27 Stephen Switzer, *Ichnographia Rustica* (1742), vol. 3, Appendix.
28 Pope to Caryll, 2 April 1722. *Alexander Pope Selected Letters*, ed. Howard Erskine-Hill (2000).

29 Mavis Batey, *Alexander Pope: The Poet and the Landscape* (1991), p. 21.
30 Horace Walpole, 'The history of modern gardening' in *Anecdotes of Painting*, vol. 4, p. 264 (Dallaway edition).
31 *The Spectator* No. 37, 12 April 1711.
32 Switzer, *Ichnographia Rustica*, vol. 3.
33 Egmont Diary (1734), Hist. MSS Comm. II, p. 138.
34 The Surrey Hills was one of the first landscapes to be designated an Area of Outstanding National Beauty in 1958.
35 Switzer, *Ichnographia Rustica* (1718), p. xxv.

FIG.8.20: *WOTTON, IN SURREY, THE SEAT OF JOHN EVELYN ESQ.*
FROM *MEMOIRS ILLUSTRATIVE OF THE LIFE AND WRITINGS OF JOHN EVELYN, ESQ. FRS,* EDITED BY WILLIAM BRAY, ESQ. (LONDON, 1819)

LEITH HILL TOWER IS SEEN IN THE DISTANCE

MAVIS BATEY, MBE, FSA, VMM, was Honorary Secretary of The Garden History Society 1971-85, President 1985-2000 and is now a Vice-President. From 1970-92 she was a part-time tutor in the Oxford Department of External Studies. She was a pioneer in garden conservation and has been a member of a number of landscape advisory committees including the English Heritage Gardens Committee. As a historian of landscape and literature, her books include *Oxford Gardens* (1982), *Historic Gardens of Oxford and Cambridge* (1989), *The English Garden Tour* (1990) with David Lambert, *Arcadian Thames* (1994), *Regency Gardens* (1995), *Jane Austen and the English Landscape* (1996), *Alexander Pope, the poet and the landscape* (1999) and *Indignation! The Campaign for Conservation* (2000).

MEMOIRS,

ILLUSTRATIVE OF

THE LIFE AND WRITINGS

OF

JOHN EVELYN, ESQ. F.R.S.

AUTHOR OF THE "SYLVA," &c. &c.

COMPRISING HIS DIARY, FROM THE YEAR 1641 TO 1705-6,

AND A SELECTION OF HIS FAMILIAR LETTERS.

TO WHICH IS SUBJOINED,

The Private Correspondence

BETWEEN

KING CHARLES I. AND HIS SECRETARY OF STATE,
SIR EDWARD NICHOLAS,

WHILST HIS MAJESTY WAS IN SCOTLAND, 1641, AND AT OTHER TIMES DURING THE CIVIL WAR;

ALSO BETWEEN

SIR EDWARD HYDE, AFTERWARDS EARL OF CLARENDON,
AND SIR RICHARD BROWNE,

AMBASSADOR TO THE COURT OF FRANCE, IN THE TIME OF KING CHARLES I. AND THE USURPATION.

THE WHOLE NOW FIRST PUBLISHED FROM THE ORIGINAL MSS.

IN TWO VOLUMES.

EDITED BY WILLIAM BRAY, ESQ.

FELLOW AND TREASURER OF THE SOCIETY OF ANTIQUARIES OF LONDON.

SECOND EDITION.—VOL. I.

LONDON: PRINTED FOR HENRY COLBURN, CONDUIT STREET;
AND SOLD BY JOHN AND ARTHUR ARCH, CORNHILL.
1819.

THE LIFE AND TIMES OF JOHN EVELYN

1620 John Evelyn, the younger son of Richard Evelyn, who made a fortune from the manufacture of gunpowder, was 'woodborn' at Wotton in the Tillingbourne Valley, which, although he spent little of his long life there, was his 'most cherished place on earth' and a great influence on him.

1625 Removed to Lewes to avoid plague then raging in London. He refused to leave an indulgent grandmother for Eton and attended the local Grammar school. His mother died at Wotton in 1635.

1637 Admitted to the Middle Temple. Went up to Balliol College, Oxford, as a Fellow Commoner; his studies were desultory and he left without a degree.

1640 Evelyn's father died and his brother George inherited Wotton. At his coming of age John was 'left to his own conduct' to face 'the greatest hazard that ever the youth of England saw'.

1641 John decided to absent himself from 'this ill face of things at home' and left for Holland. His eyes were opened to the art and architecture of the Continent. He met Lord Arundel, his neighbour at Albury, in Ghent and travelled home with him. When he returned the Civil War was imminent.

1642 He joined the King's Army for three days but, fearing that espousing the Royalist cause might prejudice his brother's estate which was in parliamentary territory, he left.

1643 Back at Wotton he resolved to 'possesse myself in some quiet if it might be, in a time of great jealosy'. His brother allowed him to make 'a triangular Pond or little stew, with an artificial rock' and a little study 'over a Cascade to passe my Malencholy houres shaded there with Trees'. Disregarding his brother's advice to 'think on your own country', in the autumn he obtained a royal licence to travel to complete his education and went to France, where he began his informative years of travel. In Paris he made a visit to the resident ambassador, Sir Richard Browne, where he met other royalist exiles.

1644 Visited and admired Pierre Morin's garden in Paris. Travelled extensively in France and Italy. In Padua he met up with Lord Arundel, who took him sight-seeing. He was the greatest art collector amongst the English nobility and influenced Evelyn at the impressionable prime of youth.

1646 He returned to Paris to stay with Sir Richard Browne, in whose library he was able to settle down to pursue his virtuoso studies; he also fell in love with his 13 year old daughter.

1647 Marriage took place in June. John composed *Instructions Oeconomiques* for his young wife. He also began to write an account of the *State of France.* In September, after nearly four years away, he left for England to settle his affairs, leaving Mary with her mother. He went to see Sayes Court, Deptford, the home of his wife's maternal uncle. The Second Civil War broke out with the royalists encamped nearby.

1649 Execution of Charles I. Evelyn kept the day as a fast. Provisionally established in a part of Sayes Court. Returned to Paris. Paid his respects to Henrietta Maria with the French Court at St Germain.

1651 After a short visit to England the previous year, Evelyn was back in Paris for the news of Charles II's defeat and with no foreseeable hope for the restoration of the monarchy, he decided to return to Sayes Court and sit out the Commonwealth in passive resistance. He was now ready to settle domestically 'having now run about the World, most part out of my owne country neere 10 years'.

1652 First child born. His *State of France* published.

1653 Although Sayes Court was sequestered he managed to obtain a lease and planned an oval garden, similar to one he had seen in Paris which was 'the beginning of all the succeeding Gardens, Walkes, Groves, Enclosures and Plantations there'.

1654 John's encounter with the Wadham set in Commonwealth Oxford had a lasting influence. Wadham's garden with ingenious experiments much impressed him. He was given one of Wren's transparent beehives to take back for his own garden. He became acquainted with many distinguished experimental philosophers, including Robert Boyle, with whom he would be reunited in Royal Society days. Thomas Sprat, one of Oxford's 'philosophical minds', recalled later, when writing his history of the Royal Society, that in Oxford they could breathe a freer air without being engaged in the 'passions and madness of that dismal age' and that 'by this Means there was a race of young men provided against the next Age'. Evelyn was with them in spirit and kept in communication throughout the Commonwealth.

1655 Mary bore another son, the only one to survive childhood. The Evelyns were now in full possession of Sayes Court, no longer sharing it with Mary's uncle. The garden he had been planning would now become a major preoccupation with groves of trees and an 'elaboratory' opening on to a flower and herb garden, a suitable place for Wren's transparent apiaries. He visited Samuel Hartlib to discuss bee-keeping with him. In spite of their radical ideas and millennial beliefs, Evelyn considered himself part of his circle, one of whom was John Beale of Herefordshire who interested Evelyn in his ideas of 'antique gardening', where attention was to be paid to the 'luckiness of situation' and good husbandry.

1656 Published a translation of *De Rerum Natura* by Lucretius.

1658 The year began tragically with the loss of his son Richard, but ended on a triumphant note when John went to London on 23 November to see the 'superb Funerall' of Cromwell, the 'Usurper'. He published *The French Gardiner,* a translation from the original by Nicolas de Bonnefons.

1659 Cromwell's son was appointed the new Protector but abdicated in May and thoughts turned to the possible restoration of the monarchy; this could not come quickly enough for John Evelyn. He already had in manuscript his life's work his 'Elysium Britannicum, or The Royal Gardens', which in the Interregnum was 'provided against the next Age'. He discussed it with Hartlib, Beale, Sir Thomas Hanmer and Sir Thomas Browne and by January a synopsis had been sent to Dr Needham for circulation to Oxford Heads of colleges, doctors, and those with gardening interests, including the Bobarts, father and son. An enthusiastic reply was received; after years devoted to puritan husbandry the dons seemed particularly keen to learn about 'the Italian gloryes and pompous beauties' of gardens which Evelyn had seen on his travels. The 'Elysium Britannicum' was a monumental undertaking and only parts of it were published in his lifetime. Since the 1960s the manuscripts had been accessible at Christ Church to scholars and are now deposited in the British Library, finally to be published by John Ingram in 2001.

1660 On 29 May Evelyn witnessed the triumphant return of Charles II. His father-in-law was also able to return after his long exile. Now secure at Sayes Court with contacts to exiled royalty made in Paris he was approaching the height of his reputation and influence. Evelyn had already kissed hands at St Germain and was gratified that when he was now presented to King Charles he called him an 'old Aquaintance'. Evelyn began to write his *Kalendarium,* which would provide the foundation of the main source of the *Diaries,* which would bring him his greatest fame.

1661 The Royal Society was founded, housed in Gresham College, and Evelyn was nominated by the king as one of its council. His good friend

Dr Wilkins was elected Chairman. *Fumifugium* 'my Book inveing against the nuisance of the Smoke of London' was published and discussed with Charles II.

1662 His 'Discourse Concerning Forest Trees' given to the Royal Society. Published *Sculptura.* With memorable pageantry at Court following the arrival of Queen Catherine of Braganza and the Queen Mother's visit to Sayes Court, Evelyn taking stock in his diary thanked God for a good year.

1664 Published his Discourse as *Sylva,* which greatly influenced landowners, now eager to improve their estates after the depletion of timber in Commonwealth days; it was the Royal Society's first official publication. Following up the theory, he 'planted the homefield and West field about Says Court' with 'Elmes, being the same Yeare that the Elmes were also planted by his majestie in Greenewich park'. He was taken by Lord Cornbury, Clarendon's son, to help in planting Cornbury Park in Oxfordshire. On the way back they stayed in Oxford, inspecting the Sheldonian theatre, of which Wren showed Evelyn the model, 'not disdaining his advice'. Publication of *Kalendarium Hortense: or The Gard'ners Almanack* as a guide to the gardener's tasks, month by month, based on experience; it was dedicated to his good friend Abraham Cowley, who had also found solace in making a garden in times of stress. He dedicated his poem 'The Garden' to Evelyn, who included one of his Latin hortulan poems in the *Elysium.*

1665 Dutch War. Evelyn appointed Commissioner to wounded and prisoners of war for the duration. Wrote an *Account of Architects and Architecture* in praise of classical architecture. In the year of the Great Plague Evelyn sent his family to Wotton but remained in London to fulfil his official duties.

1666 The Great Fire. Charles II discussed with him a 'plot for a new City'. Wren already had one.

1667 Accompanied Henry Howard, Lord Arundel's son, to 'his villa at Alburie' where I designed for him the plot for his canal and garden, with a Crypta thro the hill &c'.

1668 He translated Freart's *An Idea of the Perfection of Painting.*

1669 Evelyn went to Oxford to attend the first Encaenia in Wren's new Sheldonian theatre and to receive an honorary doctorate. He had persuaded Lord Arundel's heir to present the Arundelian antique inscriptions to Oxford University to be installed on the walls of the theatre precinct.

1671 Appointed by the King to the Council of Foreign Plantations at £500 a year.

1672 The Council now included Trade in its functions; its secretary was John Locke, according to Evelyn 'an excellent learned gentleman'. Evelyn himself was appointed Secretary of the Royal Society.

1674 Published *The Navigation and Commerce, Their Original and Progress.*

1676 Published *A Philosophical Discours of Earth.*

1685 Death of Charles II. Evelyn appointed Commissioner of the Privy Seal despite his reservations.

1688 After three years of misrule James II fled the country with his family avoiding the confrontation Evelyn feared. 'Wonderful acclamation' when William and Mary were proclaimed king and queen.

1689 Evelyn sat for his picture by Kneller for his good friend Samuel Pepys 'holding my *Sylva* in my right hand'; it was given pride of place in Pepys's library.

1691 On 22 May he received the news that transformed his life. His brother's only son died aged 35 and he became the heir to Wotton.

1694 Evelyn returned to Wotton to 'the Place that gave me a Birth and most shortly Burial' and begged his brother to let him have supervision of the garden that meant so much to him.

1697 Published *Numismata*, a treatise on medals.

1698 The Evelyns' one remaining son died, leaving one grandson, Jack still at Oxford, as heir to Wotton.

1699 On 4 October George Evelyn died and John began to repair the damage to the whole estate. Publication of *Acetaria*, a fragment of the unpublished 'Elysium Britannicum'.

1701 As Treasurer of the Royal Naval Hospital he presented the building accounts to Parliament.

1703 A great tempest, whose 'dismal groans' were heard in the Tillingbourne valley, raged in the South. New edition of *Sylva* planned. In its Preface Evelyn would claim that millions of trees had been planted after the losses in the Civil War as a result of his book and hoped that it would now influence landowners to plant following a natural disaster.

1704 *Memoires for my Grand-son.* Jack, when squire of Wotton, should be decently hospitable, but without ostentation, and not waste money on building or over-rich furniture, but invest in 'gardening, groves and walkes', and in planting oak, ash and elm on the estate.

1706 John Evelyn died in London and was buried at Wotton. His devoted wife joined him three years later. Jack, who became Sir John Evelyn in 1713, inherited and faithfully carried out the trust his grandfather had imposed on him. He was famous for his regeneration and landscaping of the Wotton region which became known as The Evelyn Country.

APPENDIX

Mr. EVELYN'S *Letter to* Mr. AUBREY

SIR,

With incredible Satisfaction, I have perus'd your *Natural History* of *Surrey*, &c. and greatly admire both your Industry in undertaking so profitable a Work, and your Judgment in the several observations which you have made. It is so useful a Piece, and so obliging, that I cannot sufficiently applaud it. Something I would contribute to it, if it were possible; but your *Spicilegium* is so accurate, that you have left nothing almost for those who shall come after you. Surrey is the Country of my Birth, and my Delight; but my Education has been so little in it, by Reason of several Accidents, that I am asham'd to discover how ignorant I am of a thousand of those excellent Remarks which I find you have taken Notice of, to my Reproach.

You have been pleas'd to mention *Wotton* (the Seat of my Brother) inviron'd as it is with Wood (from whence it takes its Denomination) and Water, and that from different Sources, capable of furnishing all the Amœnities of a *Villa*, and Garden, after the *Italian* Manner, as running 50 Foot higher than the *Arëa* of the first *Parterre*. That which I would observe to you from the Wood, is, that where goodly *Oaks* grew, and were cut down by my Grand-father almost a hundred Years since, are now altogether *Beech*; and where my Brother has extirpated the *Beech*, there rises *Birch*: Under the *Beech* spring up innumerable *Hollies*, which growing thick and close together in one of the Woods next the Meadow, is a *Viretum* all the Year long, which is a very beautiful Sight when the Leaves of the taller Trees are fallen.

It is in my *Sylva* where I give the Dimensions of a Plank of prodigious Amplitude, cut from an *Oak* growing in one of the Parks there about, or near that House, which hold almost six Foot in Breadth, and about ten in Length, half a Food in Thickness, as it remains supported on a Frame of Brick-Work. There are in the Skirts of this Parish, (which extends almost as far as the Wild of *Sussex*) certain Pits, out of which they dig *Jeate*. The Stone about the Grounds in other Parts, in the *Rag*, and what you call *Iron-Stone*, of which there lies Abundance loose in the Sands, and

JOHN AUBREY, ESQ. FRS

about certain Sugar-Loaf Mountains *South-West* of *Wotton*; which, with the Boscage upon them, and little Torrents between, make such a Solitude, as I have never seen any Place more horridly agreeable and Romatick. In the Churchyard at Wotton, digging to enlarge the Vault where our Family lies interr'd, was found an entire *Skeleton* of Gigantick Stature; it is not yet twenty Years since; but after the Work-men and Labourers had done wondering at it, and taken Measure of divers of the Bones, *&c.* (which, tho' I have not at present, I can recover from an ingenious Servant of my Brother's) without farther Curiosity, they flung into the foundation they were digging, and superstructed upon them.

In this Parish, upon *White-Down* (which is contiguous to that Tract of hills which runs from *Darking* towards *Guildford*, and so to *Portsmouth*, &c) in the Chalk-Delves, is frequently found *Cockle-Shells, Periwinckles*, &c. and in the Cart-Roats, where the Rains have gull'd, that Kind of *Pyrites* which the Country-People call *Thunder-Stones*. It is incredible what goodly *Beeches* grow upon that Hill, expos'd as they are to the most impetuous Winds, and with a very little Earth, and that extreamly loose, adhering to their Roots. From hence is one of the largest Prospects in *England*; but superior to this, is another about two Miles *South* of my Brother's House; from the Summit whereof, in a clear Day, may be seen (besides the whole Vale of Wild of *Sussex*, and much of *Kent*) part of *Eleven* other Shires; so as for the Extent and Circumference of *Vista*, I take it to be much beyond that from the *Keepe* at *Windsor*, or any that I have ever observ'd either in *England*, or elsewhere. The accent to it, is yet Northward, almost upon an even Line from the Foot of *White-Downe*. The Brow from whence this Prospect is beheld with little Aclivity, (caused by the sliding some Parts of it into the Grounds below, either by its own Weight, or some Earthquake) goes descending nine or ten Miles, almost as far as *Horseham* in *Sussex*; and the bare places from whence the Earth is slid, I have seen as far as Lewes, thirty Miles from it.

From *The Natural History and Antiquities of Surrey* by John Aubrey

SELECTED BIBLIOGRAPHY

MANUSCRIPT SOURCES

EVELYN PAPERS in the British Library, especially the following:

Add. MS 78291: Includes letters to George Evelyn from his brother, John

Add. MS 78295: Correspondence of Mary Evelyn of Wotton, eldest daughter of George Evelyn

Add. MS 78298: Letterbook of John Evelyn

Add. MS 78303: Letters to John Evelyn from his elder brother George

Add. MS 78339: 'Trades. Secrets & Receipts Mechanical, as they came casualy to hand ...' (1650s)

Add. MS 78346: 'Coleum sanitatis, or, a Particular Vegetable & Animal Dissolvant'. Autograph copy of translation

Add. MS 78349: 'Directions for the Gardiner at Says Court: But which may be of Use for other Gardens'; 1687

Add. MS 78431: Letters to Mary Evelyn from her father, Sir Richard Browne; and letters to Mary Evelyn from her husband, John Evelyn

Add. MS 78470: Includes letters to Lady Evelyn from Henrietta Pelham Holles, wife of Thomas, Duke of Newcastle.

Add. MS 78482: Letters to Sir John Evelyn from Thomas Bedingfield, steward of Wotton; 1706-35

Add. MSS 78483-4: Correspondence of Sir John Evelyn with Robert Wake, steward of Wotton; 1710-16

Add. MSS 78485-90: Correspondence of Sir John Evelyn with Thomas James, steward at Wotton; 1721-37

Add. MSS 78514B & E: Journal of Sir John Evelyn, 1st Bart.

Add. MS 78528: General account-books kept by Sir John Evelyn

Add. MS 78610: Drawings, engravings and plans, chiefly of Wotton

Add. MS 78632: Catalogue of the Evelyn library [1687?]

EGERTON MS 2540: Letters addressed to Sir John Nicholas and his son Edward Nicholas

PRIMARY PRINTED SOURCES

ADDISON, JOSEPH, *The Spectator*, no. 414, 25 June 1712

ARNOLD, MATTHEW, *Letters* (London, 1895)

AUBREY, JOHN, *The Natural History and Antiquities of the County of Surrey* (1718-9)

BOYLE, ROBERT, *The Correspondence of Robert Boyle*, ed. Michael Hunter (London, 2001)

BROWNE, SIR THOMAS, *The Letters of Sir Thomas Browne*, ed. Geoffrey Keynes (Faber & Faber, 1931, reissued 1946)

BURTON, ROBERT, *The Anatomy of Melancholy* (1621)

CAMDEN, WILLIAM, *Britannia*, ed. Edmund Gibson (1695)

CAUS, SALOMON DE, *Les Raisons des Forces mouvantes* (1615)

DEFOE, DANIEL, *The Storm, or a Collection of the Most Remarkable Casualties and Disasters* (1704)

DENNIS, JOHN, *Original Letters* (1721)

EVELYN, JOHN, *Fumifugium: or the Inconvenience of the Aer and Smoak of London Dissipated* (1661)

—— *Sculptura* (1662)

—— *Sylva: or a Discourse of Forest-Trees* (1664), 2nd edition (1670), 3rd edition (1679)

—— *Terra, a Philosophical Discourse of Earth* (1676)

—— *Kalendarium Hortense* (London: 1691 and 1699)

—— *The Compleat Gard'ner* (1693)

—— *Numismata* (1697)

—— *Acetaria* (1699)

—— *Silva: or a Discourse of Forest Trees*, 4th edition (1706)

—— *Silva: or a Discourse of Forest-Trees*, 3rd edition with Notes by A. Hunter (York, 1776 & in 2 vols. 1801)

—— *The Miscellaneous Works of John Evelyn*, ed. William Upcott (London, 1825)

—— *The Diary and Correspondence of John Evelyn*, ed. William Bray, 4 vols. (London, 1859)

—— *Sylva* (1706), ed. John Nisbet, 2 vols. (London, 1908)

—— *Memoires for my Grand-son*, ed. Geoffrey Keynes (Nonsuch edition, 1926)

—— Evelyn, John, *Directions for the Gardiner at Says-Court*, ed. Geoffrey Keynes (Nonsuch Press, 1931)

—— *The Diary of John Evelyn*, ed. Esmond S. de Beer (6 vols, Oxford, Clarendon Press, 1955)

—— & PEPYS, SAMUEL, *Particular Friends: The Correspondence of Samuel Pepys and John Evelyn*, ed. Bédoyère, Guy de la (Woodbridge, 1997)

—— *Elysium Britannicum or The Royal Gardens*, ed. John E. Ingram (Philadelphia, University of Pennsylvania Press, 2001)

GILPIN, WILLIAM, *Dialogue upon the Gardens of the Right Honourable The Lord Viscount Cobham, at Stow in Buckinghamshire* (London, 1748)

—— *Observations on the River Wye and several parts of South Wales, etc. relative chiefly to Picturesque Beauty; made in the summer of the year 1770* (London, 1782)

HARTLIB, SAMUEL, *Samuel Hartlib his legacy of husbandry* (1655), Hartlib Papers, University of Sheffield

LAWSON, WILLIAM, *A New Orchard and Garden* (1618)

LOCKE, JOHN, *Essay Concerning Human Understanding* (1690)

MORE, HENRY, *The Immortality of the Soul* (1658)

MORE, THOMAS, *Utopia* (1516)

PEACHAM, HENRY, *The Compleat Gentleman*

PEPYS, SAMUEL, *The Diary of Samuel Pepys*, ed. R. Latham and W. Matthews (London, 1971-1995)

PLOT, ROBERT, *The Natural History of Oxfordshire* (1675)

POPE, ALEXANDER, *Alexander Pope Selected Letters*, ed. Howard Erskine-Hill (2000)

QUINTINYE, JEAN DE LA, *Le Parfait Jardinier ou Instruction pour les Jardins Fruitiers et Potagers* (1690)

SHARROCK, ROBERT, *The History of the Propagation and Improvement of vegetables by the concurrence of Arte and Nature, written according to Observations made from Experience and Practice* (1659)

STELLUTI, FRANCESCO, *Trattato del Legno fossile minerale* (1637)

SWITZER, STEPHEN, *Ichnographia Rustica* (1718 & 1742)

TYSON, EDWARD, *Orang-Outang, sive Homo Sylvestris: or the Anatomy of a Pygmie Compared with that of a Monkey, and Ape and a Man* (1699)

WALPOLE, HORACE, *Catalogue of Engravers* (1763)

WORTHINGTON, JOHN, *The Diary and Correspondence of Dr John Worthington*

WOTTON, HENRY, *The Elements of Architecture* (1624)

YOUNG, ARTHUR, *Annals of Agriculture*, vol. 22 (1794)

SECONDARY SOURCES

BALDOCCHI, DENNIS, 'The carbon cycle under stress', *Nature*, vol. 437, 22 September 2005

BATEY, MAVIS, *Oxford Gardens* (1982)

—— *Alexander Pope: The Poet and the Landscape* (1991)

—— 'The Pleasures of the Imagination: Joseph Addison's Influence on Early Landscape Gardens', *Garden History*, vol. 33, no. 2 (2005)

BELLAMY, PAT H. *et al.*, 'Carbon losses from all soils across England and Wales 1978-2003', *Nature*, vol. 437, 8 September 2005

BOWLE, JOHN, *John Evelyn and his World* (London, 1981)

BRANDON, PETER, 'Land, Technology and Water Management in the Tillingbourne Valley', *Southern History*, vol. 6 (1984)

—— *The Story of the Tillingbourne* (Shere, Surrey, 1984 and later editions)

—— *The North Downs* (Chichester, 2005)

BRAYLEY, E.W., *A Topographical History of Surrey*, ed. Edward Walford (1848)

BRAYNE, MARTIN, *The Greatest Storm* (Thrupp, Gloucestershire, Sutton Publishing, 2002)

BRENT, COLIN, *Pre-Georgian Lewes: c. 890-1714* (Lewes, 2004)

BRITTON, JOHN (ed.), *The Natural History of Wiltshire* (1847)

CAMPBELL-CULVER, MAGGIE, *A Passion for Trees: the Legacy of John Evelyn* (London, 2006)

CHAMBERS, DOUGLAS, 'John Evelyn and the Invention of the Heated Greenhouse', *Garden History*, vol. 20, no. 2 (1992)

—— ' "Elysium Britannicum not printed neere ready &c.": The "Elysium Britannicum" in the Correspondence of John Evelyn', in Therese O'Malley and Joachim Wolschke-Bulmahn (eds.), *John Evelyn's 'Elysium Britannicum' and European Gardening (Dumbarton Oaks,* 1998)

CHANEY, EDWARD, 'Evelyn, Inigo Jones, and the Collector Earl of Arundel', in *John Evelyn and his Milieu* (London, British Library, 2003)

CHIGNELL, ROBERT, *The Life and Paintings of Vicat Cole, RA* (London, 1898)

CIAIS, PH. et *al.*, 'Europe-wide reduction in primary productivity caused by the heat and drought in 2003', *Nature*, vol. 437, 22 September 2005

COLVIN, HOWARD, 'The South Front of Wilton House', *Essays in English Architectural History* (Yale University Press, 1999)

DARLEY, GILLIAN, *John Evelyn: Living for Ingenuity* (Yale University Press, 2006)

—— 'John Evelyn's Norwich Garden', *Garden History*, vol. 32, no. 2 (2006)

FREEDBERG, DAVID, *The Eye of the Lynx; Galileo, His Friends, and the Beginning of Modern Natural History* (Chicago, the University of Chicago Press, 2002)

GIBSON-WOOD, CAROL, 'Susanna and Her Elders: John Evelyn's Artistic Daughter', in *John Evelyn and His Milieu* (London, British Library, 2003)

GOODCHILD, PETER, 'No Phantasticall Utopia, but a Reall Place. John Evelyn, John Beale and Backbury Hill', *Garden History*, vol. 19, no. 2 (1991)

GRIFFITHS, ANTONY, 'The Etchings of John Evelyn', in David Howarth (ed.), *Art and Patronage in the Caroline Courts* (Cambridge University Press, 1993)

—— 'John Evelyn and the Print', in *John Evelyn and his Milieu* (London, British Library, 2003)

HARRIS, FRANCES, *Transformations of Love: the Friendship of John Evelyn and Margaret Godolphin* (Oxford University Press, 2003)

HARRIS, FRANCES & HUNTER, MICHAEL (eds.), *John Evelyn and his Milieu* (London, British Library, 2003)

HISCOCK, W.G., *John Evelyn and his Family Circle* (London, 1955)

HUNT, JOHN DIXON, *Garden and Grove* (1986)

HUNT, JOHN DIXON AND WILLIS, PETER (eds.), *The Genius of the Place: The English Landscape Garden 1620-1820* (Paul Elek, 1975)

JACQUES, DAVID & HORST, AREND JAN VAN DER, *The Gardens of William and Mary* (London, Christopher Helm, 1988)

LAIRD, MARK, 'Parterre, Grove, and Flower Garden: European Horticulture and Planting Design in John Evelyn's Time', in Therese O'Malley and Joachim Wolschke-Bulmahn (eds.), *John Evelyn's 'Elysium Britannicum' and European Gardening* (Dumbarton Oaks, 1998)

—— 'Sayes Court Revisited', in *John Evelyn and His Milieu* (London, British Library, 2003)

—— 'Perpetual Spring or Tempestuous Fall: the Greenhouse and the Great Storm of 1703 in the Life of John Evelyn and his Contemporaries', *Garden History*, vol. 34, no. 2 (2006)

LAMBIN, DENIS, 'Pleasure with profit: some aspects of landscape gardening', *Garden History*, vol. 3, no. 2 (1975)

LEITH-ROSS, PRUDENCE, 'A Seventeenth-century Paris Garden', *Garden History*, vol. 21, no. 2 (1993)

—— 'The Garden of John Evelyn at Deptford', *Garden History*, vol. 25, no. 2 (1997)

—— 'Fruit Planted around a New Bowling Green at John Evelyn's Garden at Sayes Court, Deptford, Kent in 1684/5', *Garden History*, vol. 31, no. 1 (2003)

MANDELBROTE, GILES, 'John Evelyn and his Books', in *John Evelyn and his Milieu* (London, British Library, 2003)

MORRIS, CHRISTOPHER (ed.), *The Illustrated Journeys of Celia Feinnes 1685–c.1712* (Macdonald, 1982)

OGDEN, H.V. & M.S., *English Taste in Landscape in the Seventeenth Century* (1955)

PINCUS, STEVEN, 'John Evelyn: Revolutionary', in *John Evelyn and his Milieu* (London, British Library, 2003)

ROBINSON, JOHN MARTIN (ed.), *Remembrances of Things Worth Seeing in Italy Given to John Evelyn 25 April 1646 by Thomas Howard, 14th Earl of Arundel* (Roxburghe Club, 1987)

ROHDE, ELEANOR SINCLAIR, *Oxford's College Gardens* (Jenkins, 1932)

SCHIERMEIER, QUIRIN, 'Hurricane link to climate change is hazy', *Nature*, vol. 437, 22 September 2005

SCULL, G.D., *The Evelyns in America* (1881)

SMALL, CAROLA & ALASTAIR, 'Evelyn and the Garden of Epicurus', *Journal of the Warburg and Courtauld Institute*, vol. 60 (1997)

SPENSER, EDMUND, *The Faeirie Queene*

STRONG, ROY, *The Renaissance Garden in England* (1979)

SULLIVAN, ISABEL, 'Their "Own Sweet Country": the Evelyns in Surrey', in *John Evelyn and his Milieu* (London, British Library, 2003)

WALPOLE, HORACE, 'The history of modern gardening', in *Anecdotes of Painting*, vol. 4 (Dallaway edition)

WIJNANDS, D.O., 'Hortus auriaci: the gardens of Orange and their place in late seventeenth-century botany and horticulture', *Garden History*, vol. 8, nos. 2 & 3 (1998)

WOODS, MAY & WARREN, ARETE SWARTZ, *Glass Houses: A History of Greenhouses, Orangeries and Conservatories* (London, Aurum Press, 1899)

ZWAAN, MARISCA SIKKENS-DE, 'Magdalena Poulle (1632-99): A Dutch Lady in the Circle of Botanical Collectors', *Garden History*, vol. 30, no. 2 (2002)

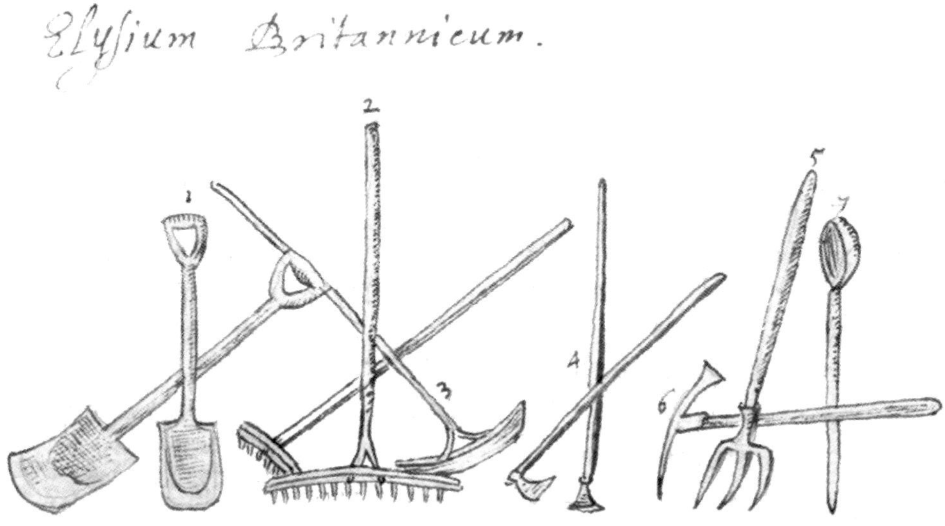

LIST OF ILLUSTRATIONS

Abbreviations used in the credits: BL (By permission of The British Library), RHS (©Royal Horticultural Society, Lindley Library), SHS (Reproduced by permission of Surrey History Service, copyright of Surrey History Service), *Memoirs*, ed. Bray (*Memoirs illustrative of the Life and Writings of John Evelyn, Esq. F.R.S.*, ed. William Bray (London 1819))

INDEX

This index covers the text of chapters 1-8 and all the illustrations but excludes endnotes. Where book titles have been abbreviated, the full title may be found in the Bibliography. References to illustrations and their captions are indicated by **bold** type. The following abbreviations have been used: JE = John Evelyn, the diarist; Jack = Sir John Evelyn, grandson of JE.

The book contains many references to and quotations from JE's diary. Only those mentioned in the text are included in the index.